KT-385-628

TUTORIAL CHEMISTRY TEXTS

21
Inorganic Chemistry in Aqueous Solution

JACK BARRETT

formerly of
Imperial College of Science, Technology and Medicine,
University of London

RS•C
ROYAL SOCIETY OF CHEMISTRY

Cover images © Murray Robertson/visual elements 1998–99, taken from the
109 Visual Elements Periodic Table, available at www.chemsoc.org/viselements

ISBN 0-85404-471-X

A catalogue record for this book is available from the British Library

Published by The Royal Society of Chemistry, Thomas Graham House, Science Park,
Milton Road, Cambridge CB4 0WF, UK
Registered Charity No. 207890
For further information see our web site at www.rsc.org

Typeset in Great Britain by Alden Bookset, Northampton
Printed and bound by Italy by Rotolito Lombarda

Preface

Water is an important molecule. Seventy-one percent of the Earth's surface is covered by liquid water, it is the most abundant molecule, and it is a good solvent for polar and ionic substances. Samuel Taylor Coleridge's Ancient Mariner knew this:

Water, water every where,
Nor any drop to drink.

The lack of pure water for drinking purposes and the lack of water sufficiently pure for irrigating the land are major challenges for the human race.

This book offers no solutions to such severe problems. It consists of a review of the inorganic chemistry of the elements in all their oxidation states in an aqueous environment. Chapters 1 and 2 deal with the properties of liquid water and the hydration of ions. Acids and bases, hydrolysis and solubility are the main topics of Chapter 3. Chapters 4 and 5 deal with aspects of ionic form and stability in aqueous conditions. Chapters 6 (s- and p-block), 7 (d-block) and 8 (f-block) represent a survey of the aqueous chemistry of the elements of the Periodic Table. The chapters from 4 to 8 could form a separate course in the study of the periodicity of the chemistry of the elements in aqueous solution, chapters 4 and 5 giving the necessary thermodynamic background. A more extensive course, or possibly a second course, would include the very detailed treatment of enthalpies and entropies of hydration of ions, acids and bases, hydrolysis and solubility.

There are many tables of data in the text and the author has spent much time in attempting to ensure maximum consistency with the various available sources.

I thank Martyn Berry for reading the manuscript and for his many suggestions that led to improvements in the text.

Jack Barrett
Kingston-upon-Thames

TUTORIAL CHEMISTRY TEXTS

EDITOR-IN-CHIEF

Professor E W Abel

EXECUTIVE EDITORS

Professor A G Davies
Professor D Phillips
Professor J D Woollins

EDUCATIONAL CONSULTANT

Mr M Berry

This series of books consists of short, single-topic or modular texts, concentrating on the fundamental areas of chemistry taught in undergraduate science courses. Each book provides a concise account of the basic principles underlying a given subject, embodying an independent-learning philosophy and including worked examples. The one topic, one book approach ensures that the series is adaptable to chemistry courses across a variety of institutions.

Further information about this series is available at www.rsc.org/tct

Order and enquiries should be sent to:
Sales and Customer Care, Royal Society of Chemistry, Thomas Graham House,
Science Park, Milton Road, Cambridge CB4 0WF, UK

Tel: +44 1223 432360; Fax: +44 1223 426017; Email: sales@rsc.org

Contents

Some Thermodynamic Symbols and Definitions Used in the Text

Changes in the state functions (enthalpy, H, Gibbs energy, G, and entropy, S) are indicated in the text by a Greek delta, Δ, followed by a subscript that indicates the type of change. The various changes and their particular subscripts are defined in the table:

Change	Symbol
Formation of a compound or ion from its elements in their standard states	Δ_f
Formation of gaseous atoms of an element from the element in its standard state	Δ_a
Formation of the gaseous form of an element or compound from the standard state	Δ_v
Formation of a solid crystal lattice from its constituent gaseous ions	Δ_{latt}
Formation of a hydrated ion from its gaseous state	Δ_{hyd}
Formation of a solution of a compound from its standard state	Δ_{sol}

Other frequently used relationships are:

$$\Delta_{ion} H^{\oplus}(M, g) = I(M, g) + 6.2$$

$$\Delta_{ea} H^{\oplus}(X, g) = E(X, g) - 6.2$$

These convert internal energy changes to enthalpy changes.
Values of some constants:

N_A (Avogadro) $= 6.022(14199) \times 10^{23}$ mol^{-1}

F (Faraday) $= 96485.(34)$ C mol^{-1}

e (electronic charge) $= 1.602(1773) \times 10^{-19}$ C

h (Planck) $= 6.626(0755) \times 10^{-34}$ J s

R (molar gas) $= 8.314(472)$ J mol^{-1} K^{-1}

The subscript ion refers to the ionization of a gaseous atom, M, and ea refers to the electron attachment to a gaseous atom, X, providing the values are quoted with units of kJ mol^{-1}. $I(M, g)$ is the ionization energy of M, and $E(X, g)$ is the energy change occurring when the atom X accepts an electron to become a negative ion: its electron attachment or electron gain energy. Other definitions and explanations of nomenclature are given as appropriate in the text.

The brackets indicate the fine accuracy of the values that are normally ignored in general calculations.

1
Water and its Solvent Properties

This introductory chapter is about liquid water and some of its important properties – those that enable it to act as a good solvent for ionic and polar substances.

Aims

By the end of this chapter you should understand:

- What is meant by the terms solute and solvent
- That the water molecule has a significant dipole moment
- That the molecules in solid and liquid water interact by hydrogen bonding and that hydrogen bonding is responsible for the anomalous properties of water compared with the other hydrides of Group 16
- That the hydrogen bonding occurs because of the large dipole moment of the water molecule, and the presence of the unshielded protons
- That the self-ionization of water is slight and is consistent with the poor electrical conductance of the liquid
- That liquid water possesses a high relative permittivity which is associated with its property as a good solvent for polar molecules and ionic compounds
- That water is thermodynamically very stable and may act as solvent for a large range of compounds

1.1 Introduction

A **solution** consists of a **solute** dissolved in a **solvent**. The solute is recoverable from the solution, *e.g.* sodium chloride dissolved in water is

Some reactions with liquid water are dangerous, that of sodium metal with water being one of them. As quoted in Martyn Berry's book *H₂O and All That*, one student wrote in an examination answer: "Sodium is so dangerous that it isn't handled by human beings at all, only by chemistry teachers".

recoverable by evaporating the solvent. In some cases the "solute" reacts with water and cannot be recovered by the removal of the solvent. In such cases the solution produced is of another solute, related to the initially dissolved substance. For example, sodium metal reacts with water to give a solution of sodium hydroxide. In general, polar molecules and ionic solids dissolve readily in polar solvents, and non-polar molecules dissolve readily in non-polar solvents. The ionic compound sodium chlorate (VII) ($NaClO_4$, sodium perchlorate) dissolves in water at 25 °C to give a **saturated solution** – one with the maximum solubility – containing 205 grams per 100 cm^3 of water.

Atomic mass units. All atomic masses are based upon that of the mass-12 isotope of carbon, ^{12}C. This is accepted to be equal to the **molar mass** of the isotope, $M(^{12}C)$ [$= 12 \times 10^{-3}$ kg mol^{-1}] divided by the Avogadro constant, N_A, *i.e.* $m(^{12}C) = M(^{12}C)/N_A$. The atomic unit of mass is 1 $m_u = m(^{12}C)/12$. The quantity $M(^{12}C)/12$ is known as the molar mass constant, M_u [$= 1 \times 10^{-3}$ kg mol^{-1}]. Relative atomic masses are symbolized as $A_r(X)$ and are equal to $m(X)/m_u$ or $M(X)/M_u$. Likewise, relative molecular masses are denoted by the symbol M_r(molecular formula) = molar mass/M_u.

Box 1.1   Solubility

The **solubility** of a solute may be expressed in different ways. It is the **concentration** of the solute in a **saturated solution** at a particular temperature (because solubility varies with temperature). A saturated solution at a given temperature is one that is in equilibrium with undissolved solute.

(i) **Molar concentration**, c (still sometimes called **molarity**, a term now forbidden by the International Union of Pure and Applied Chemistry). Applied normally to dilute solutions, this is the number of moles of the solute in 1 dm^3 (1 litre) of the solution, expressed in terms of moles per cubic decimetre: mol dm^{-3}. A typical dilute solution would be one with a concentration of 0.01 mol dm^{-3}.

(ii) **Molality**, m; this is appropriate for more concentrated solutions, and is the number of moles of the solute in 1 kg of the solvent. A solution containing 10 moles of solute in 1 kg of solvent would be 10 molal or 10 mol kg^{-1}.

(iii) **Mole fraction**, x; this is the concentration of a solute in terms of the number of moles of the solute expressed as a fraction of the numbers of moles of the solute and the solvent. A 10 molal aqueous solution would have a solute mole fraction of $10/65.509 = 0.153$, since 1 kg of water contains $1000/18.0152 = 55.509$ moles of water. [$A_r(O) = 15.9994$, $A_r(H) = 1.0079$]

Molality and mole fraction, as expressions of concentration of a solute, possess the advantage over molar concentration of being independent of temperature.

Worked Problem 1.1

Q A saturated solution of sodium chloride in water at 20 °C contains 36 g of the salt per 100 g of water. The density of solid sodium chloride is 2170 kg m^{-3} and that of the saturated solution is 1197 kg m^{-3}. The density of water at 20 °C is 998.2 kg m^{-3} [A_r(Na) = 22.99, A_r(Cl) = 35.453]

(a) Express the concentration of sodium chloride as (i) molar concentration, (ii) molality and (iii) mole fraction. (b) Calculate the change in volume that occurs when the solution is made from its constituents, and suggest an explanation for the change.

A (a) (i) The molar concentration of a saturated solution may be calculated from its density. The volume of the solution is calculated by dividing its mass by its density: $V = 0.136$ kg/ 1197 kg m^{-3} = 1.136 × 10^{-4} m^3 or 0.1136 dm^3. One dm^3 would contain 36/0.1136 = 316.9 g of NaCl. The molar concentration of the solution is 316.9/(22.99 + 35.453) = 5.422 mol dm^{-3}.
(ii) There would be 360 g NaCl dissolved in 1 kg of water. The molality of the solution is 360/(22.99 + 35.453) = 6.16 *molal*.
(iii) The mole fraction of sodium chloride in the solution is:

$$x_{NaCl} = \frac{\dfrac{36}{58.443}}{\left(\dfrac{36}{58.443} + \dfrac{100}{18.015}\right)} = 0.0999$$

(b) One dm^3 of the solution has a mass of 1197 g, of which 316.9 g are NaCl, leaving 1197 − 316.9 = 880.1 g of water. The volumes of these constituents are (316.9 × 10^{-3} kg)/(2170 kg m^{-3}) = 1.4604 × 10^{-4} m^3 = 146.04 cm^3 NaCl and (0.8801 kg)/(998.2 kg m^{-3}) = 8.8169 × 10^{-4} m^3 = 881.69 cm^3 water, making a total volume of 146.04 + 881.69 = 1027.73 cm^3. There is therefore a contraction of 27.73 cm^3 when the solution is made. This indicates that there are interactions of the sodium cations and chloride anions with the solvent as they become **hydrated**.

α-D-Glucose, $C_6H_{12}O_6$, with its five **hydrophilic** (= water seeking) OH groups, shown in Figure 1.1, dissolves readily in water to the extent of 54.6% by mass at 30 °C, and sucrose at 100 °C is soluble in water to the extent of 83% by mass.

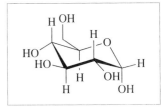

Figure 1.1 The α-D-glucose molecule

The alcohols, methanol, ethanol and 1- and 2-propanol are completely **miscible** with water, *i.e.* any composition is possible between 0% and 100% of the alcohol and 100% and 0% of water. As the **hydrophobic** (= water repelling) hydrocarbon content increases, alcohols become progressively less soluble in water, pentadecanol dissolving only to the extent of 1 × 10^{-5}% by mass and would be classed as insoluble. Non-polar tetrachloromethane dissolves in water at 25 °C to give an 8 × 10^{-3} mol dm^{-3} solution and hexane dissolves to give only a 1.3 × 10^{-4} mol dm^{-3} solution.

The hydration of ions – their interactions with water as solvent – is the subject of Chapter 2.

1.2   Liquid Water

Water is the most abundant molecular substance on Earth. The Earth's **hydrosphere** contains an estimated 1.41 × 10^{24} grams of water in all its phases, contained mainly by the oceans [97% as saline water covering

70.8% of the Earth's surface] with only 2% existing in the solid state as polar ice and glaciers. Ground water, aquifers, lakes, soil moisture and rivers account for the very small remainder. Like all liquids and solids, water exerts a vapour pressure and at any time there are about 1.3×10^{19} grams in the atmosphere (0.0009% of the Earth's total) and it is replenished every 12 days or so.[1] This amount seems to be rather small, but if all the water vapour were to be precipitated evenly over the Earth's surface instantaneously as rain there would be a layer 2.5 cm thick. The vapour is responsible for a substantial fraction of global warming, the retention of energy in the atmosphere, in the absence of which the Earth's surface would be some 33 °C cooler on average.

The triatomic water molecule has a bond angle of 104.5° in its electronic ground state, and the O—H bond lengths are 96 pm. Its structure is shown in Figure 1.2(a). The electronegativity coefficients (Allred–Rochow)[2] of hydrogen (2.1) and oxygen (3.5) are sufficiently different to make the molecule **polar** with a **dipole moment** of 1.84 D [1 Debye (D) $= 3.33564 \times 10^{-30}$ C m]. The **dipole** of the molecule is shown in Figure 1.2(b), the oxygen end being negative with respect to the two hydrogen atoms. In addition to the normal van der Waals intermolecular forces that operate between molecules, the relatively "bare" protons of the water molecule and the electronegative – and so relatively electron-rich – oxygen atom allow the formation of hydrogen bonds between adjacent molecules in the liquid and solid states. Hydrogen bonds in water have bond enthalpies of about 20 kJ mol^{-1}, which is weak compared with the strengths of single covalent bonds, which lie in the region 44 (Cs—Cs) to 570 (H—F) kJ mol^{-1}. However, H-bonds are responsible for the abnormally high values of the melting and boiling points of water, considering its low relative molar mass of 18.

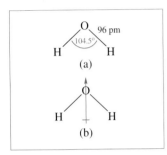

Figure 1.2 (a) The structure of the water molecule; (b) the water molecule dipole moment

Box 1.2   Dipole Moments

Polar molecules possess **electric dipole moments**, *e.g.* HF is a dipolar molecule with a partial charge of $+0.413e$ on the hydrogen atom and a partial charge of $-0.413e$ on the fluorine atom. The partial charges arise from the difference in electronegativity coefficients of the two atoms causing an unequal sharing of the valence electrons. The two partial charges are separated by a distance of 92 pm, the equilibrium internuclear distance otherwise known as the bond length. The dipole moment (normally the "electric" term is taken for granted) of the molecule, μ, is the product of the charge on the positive end, q, and the bond length, r:

$$\mu = qr \qquad (1.1)$$

The dipole moment of the HF molecule is given by:

$$\mu(\text{HF}) = 0.413 \times 1.602 \times 10^{-19}\,\text{C} \times 92 \times 10^{-12}\,\text{m} = 6.091 \times 10^{-30}\,\text{C m}$$

The very small quantity is converted into a more easily quotable one by dividing the result by 3.33564×10^{-30}, giving 1.826 D for the dipole moment in Debye units. The Debye is a non-S.I. unit related to the S.I. unit C m by the relation:

$$1\,\text{D} = 3.33564 \times 10^{-30}\,\text{C m}$$

Figure 1.3 A diagram of the water molecule showing the dipole moment and its two constituent bond dipoles

The dipole moment of the water molecule is 1.8546 D or $1.8546 \times 3.33564 \times 10^{-30} = 6.19 \times 10^{-30}$ C m. Regarding this as the resultant of two O–H bond dipole moments, as shown in Figure 1.3, the charge separation in each bond is given by:

$$q = \frac{6.19 \times 10^{-30}}{2 \times 96 \times 10^{-12} \times \cos(104.5/2) \times 1.602 \times 10^{-19}} = 0.33e$$

This indicates that the charge separation is equivalent to a partial charge of $-0.66e$ on the oxygen atom and a partial charge of $+0.33e$ on both hydrogen atoms. This is to be expected from the difference in electronegativity coefficients of the two atoms in each bond.

Van der Waals forces operate between molecules, and are classified as: (i) dipole–dipole interactions or Keesom forces, (ii) dipole–induced-dipole interactions, and (iii) London dispersion forces which operate even between pairs of atoms as their electron distributions vary at any particular instant from perfect symmetry around their nuclei. They are responsible for the cohesion displayed between atoms of Group 18 (even helium can be liquefied) and between molecular compounds, and determine their melting and boiling points to a large extent. They operate in giant arrays such as metals and covalent lattices, *e.g.* diamond, but contribute only to a small extent to the total cohesion of the solid and liquid states in those cases.

Worked Problem 1.2

Q From the data given in Table 1.1, what evidence is there for hydrogen bonding in liquid and solid water?

Table 1.1 Melting and boiling temperatures (°C) for hydrides of the elements of Groups 14 and 16

Group 16 hydride	M.p.	B.p.	Group 14 hydride	M.p.	B.p.
H_2O	0	100	CH_4	−182	−161
H_2S	−85	−61	SiH_4	−185	−112
H_2Se	−60	−41	GeH_4	−165	−89
H_2Te	−49	−2	SnH_4	−150	−52

A A comparison of melting and boiling points for the series of hydrides of Groups 14 and 16 indicates the greater cohesion in the case of solid and liquid water. As the data for the Group 14 hydrides indicate, there is a general increase of the temperatures as the relative molecular masses increase, but superimposed on the similar trend in the Group 16 hydrides is the effect of hydrogen bonding in the case of water.

Over many years, rivers have carried the results of weathering of the rocks to the oceans, which have an enormous total ionic content as indicated by the data given in Table 1.2. Typically, when 1 dm^3 of seawater is evaporated to dryness, 42.8 grams of solid are produced, which contains sodium chloride (58.9%), magnesium chloride hexahydrate [$MgCl_2.6H_2O$] (26.1%), sodium sulfate decahydrate [$Na_2SO_4.10H_2O$] (9.8%), calcium sulfate (3.2%) and potassium sulfate (2%). Other compounds are present in minute amounts.

Table 1.2   The concentrations of the main constituent elements dissolved in sea water

Element	Concentration/mg dm^{-3}
Chlorine	19,400
Sodium	10,800
Magnesium	1290
Sulfur	905
Calcium	412
Potassium	399
Bromine	67
Carbon (as carbonate and hydrogen carbonate ions)	28
Strontium	8
Boron	4.4
Silicon	2.2
Fluorine	1.3

Worked Problem 1.3

Q   Using the data given in the text, calculate the amount of sodium chloride that is potentially available from the oceans. Take the volume of the oceans to be 1.37×10^{18} m^3.

A   The amount of available sodium chloride is $0.589 \times 1.37 \times 10^{18}$ (m^3) $\times 10^3$ (dm^3 m^{-3}) $\times 42.8$ (g) $\times 10^{-3}$ (kg g^{-1}) $= 3.45 \times 10^{19}$ kg.

Currently, world annual production of the compound is about 2×10^{11} kg.

The major physical properties of water are given in Table 1.3. The abnormally high melting and boiling points already referred to are caused by hydrogen bonding in the solid and liquid phases, respectively. The structure of solid water (ice) formed at 0 °C and 100 kPa pressure, called ice-I$_h$, is shown in Figure 1.4.

h = hexagonal

Table 1.3   Major physical properties of water

Property	Value
Melting point (101 325 Pa pressure)	0 °C, 273.15 K
Boiling point (101 325 Pa pressure)	100 °C, 373.15 K
Temperature of maximum density	4 °C, 277.13 K
Maximum density	999.975 kg m^{-3}
Density at 25 °C	997.048 kg m^{-3}
Relative permittivity, ε_r, at 25 °C	78.54
Electrical conductivity at 25 °C	5.5×10^{-6} S^a m^{-1}
Ionic product [H$^+$][OH$^-$] at 25 °C, K_w	1.008×10^{-14}
Enthalpy of ionization at 25 °C	55.83 kJ mol^{-1}
Standard enthalpy of formation, $\Delta_f H°$	-285.83 kJ mol^{-1}
Standard Gibbs energy of formation, $\Delta_f G°$	-237.1 kJ mol^{-1}

^{a} 1 Siemen (S) = 1 Ω^{-1} (reciprocal ohm)

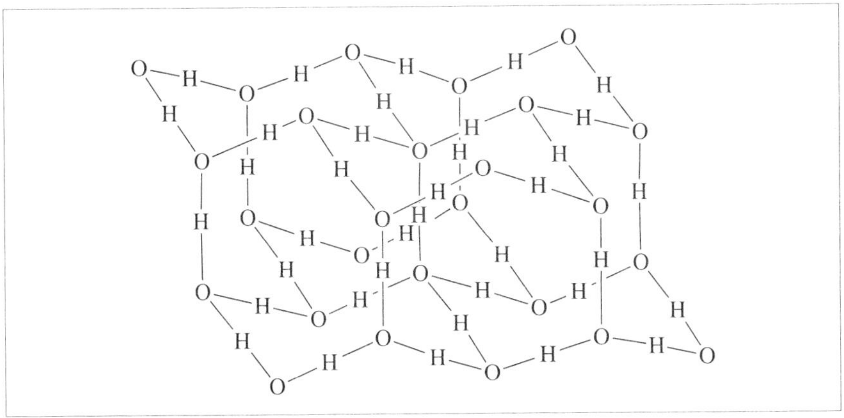

Figure 1.4   The structure of ice-I$_h$; the hydrogen atoms are placed symmetrically between the O−O pairs for simplicity

Ice-I$_h$ consists of sheets of oxygen atoms arranged in a chair-like manner, as shown in the margin, with hydrogen atoms asymmetrically placed between all the adjacent oxygen atom pairs. The sheets are linked together with O−H−O bonds. Each oxygen atom is surrounded by a nearly tetrahedral arrangement of oxygen atoms; there are three oxygen atoms at a distance of 276.5 pm (within the sheets) and a fourth oxygen atom at a distance of 275.2 pm (linking the sheets.) The arrangement of the hydrogen atoms is disordered because of their asymmetrical placement between the pairs of oxygen atoms at any one time.

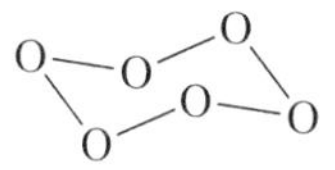

The somewhat open network structure of solid water determines that the density of ice at 0 °C is 916.7 kg m^{-3}. That of liquid water at 0 °C is 999.8 kg m^{-3} so solid ice floats on water, a fact noticed eventually by the captain of the *Titanic*! In liquid water at 0 °C there is still considerable

Natural water contains 0.15% of the mass-2 isotope of hydrogen: deuterium, ^{2}H or D. Heavy water (~100% D$_2$O) has its maximum density at 11.2 °C, an indication of the somewhat stronger deuterium bonding between adjacent molecules in the liquid phase. Solid D$_2$O melts at 3.82 °C and the liquid boils at 101.42 °C.

order because of the extensive hydrogen bonding. As the temperature rises, individual molecules have more translational, vibrational and rotational energy and need more space in which to move, thus causing most liquids (and solids) to expand and to have a lower density. This tendency is present in liquid water as the temperature increases, but additionally there is a progressive breakage of the hydrogen-bonded system that allows the open structure to collapse and to cause the density to increase.

Between 0 °C and the temperature of maximum density (4 °C) the hydrogen bond collapse dominates over the normal thermal expansion. At temperatures above that of the maximum density, thermal expansion dominates, and the density decreases progressively as the temperature rises.

The magnitude of the **relative permittivity** (or **dielectric constant**), ε_r of water is crucial to its solvent properties. In a vacuum, when two electric charges, q_1 and q_2 are brought together from an infinite distance to a separation r, the potential energy, E_p, is given by the **Coulomb's law** equation as:

$$E_p = \frac{q_1 q_2}{4\pi\varepsilon_0 r} \tag{1.2}$$

where ε_0 is the vacuum permittivity. It has a value of 8.854×10^{-12} $J^{-1} C^2 m^{-1}$.

When the same procedure takes place in a medium such as liquid water, the vacuum permittivity in equation (1.2) is replaced by the permittivity of the medium. Normally the permittivities for a variety of solvents are expressed as relative permittivities, ε_r, at given temperatures. Some typical values of relative permittivites are given in Table 1.4.

Quantities such as ε_0 and the electronic charge, e, are accurately known to several decimal places, but are in general usage restricted to three decimal places: $\varepsilon_0 = 8.854 \times 10^{-12}$ $J^{-1} C^2 m^{-1}$ and $e = 1.602 \times 10^{-19}$ C.

There are two liquids that have larger relative permittivities than water, namely anhydrous HF (84 at 0 °C) and formamide [or methanamide, $O{=}CH(NH_2)$] (109 at 20 °C), but they are very reactive to most solutes.

Table 1.4 Some typical values of relative permittivities

Compound	Temperature/ °C	Relative permittivity, ε_r
Water	25	78.54
Methanol	25	32.63
Liquid ammonia	−33.4 (b.p.)	22.4
Propanone	25	20.7
CCl$_4$	20	2.24
Benzene	20	2.28
Hexane	20	1.89

The great significance of the high value of relative permittivity of water is explored in Chapter 2.

The electrical conductance of liquid water is very low compared with the values given by solutions of ionic compounds. Typically, the conductance of a 1 mol dm^{-3} solution of sodium chloride is about one million times higher than that of water. This illustrates the effect of the dissociation of ionic substances when they are dissolved in water:

$$Na^+Cl^-(s) \rightarrow Na^+(aq) + Cl^-(aq) \qquad (1.3)$$

The ionic product of water, K_w, is related to the equilibrium:

$$H_2O(l) \rightleftharpoons H^+(aq) + OH^-(aq) \qquad (1.4)$$

in which liquid water dissociates slightly to give equal concentrations of hydrated protons and hydrated hydroxide ions. The equilibrium constant for the reaction is:

$$K = \frac{a_{H^+}\, a_{OH^-}}{a_{H_2O}} \qquad (1.5)$$

in which the a terms represent the ratios of the activities of the species shown as subscripts to those at the standard activity of 1 mol dm^{-3}. The activity of liquid water in the solution is taken to be 1, because in dilute solutions $a_{solvent} = a^\circ_{solvent}$ so the equation becomes:

$$K_w = a_{H^+}\, a_{OH^-} \qquad (1.6)$$

and is known as the autoprotolysis constant or ionic product of water.

The **standard state** of a substance (symbol $^\circ$) is its pure form (solid, liquid or gas) at a pressure of 1 bar ($= 10^2$ kPa; 1 Pa $= 1$ N m^{-2}) and at a specified temperature. If the temperature is not specified, it is assumed to be 298.15 K or 25 °C. The standard molar activity of a solute is 1 mol dm^{-3}.

In dilute solutions the activity of an ion can be defined by the equation:

$$a = \gamma c / a^\circ \qquad (1.7)$$

where c is the molar concentration (in mol dm^{-3}) of the solute, γ is the **activity coefficient** and a° is the standard molar activity of 1 mol dm^{-3}.

In very dilute solutions, γ may be taken to be 1.0 and the autoprotolysis constant may be formulated as:

$$K_w = [H^+(aq)][OH^-(aq)] \qquad (1.8)$$

the square brackets indicating the molar concentration of the substance by the usual convention. The autoprotolysis constant of water is essential for the discussion of pH and the acid/base behaviour of solutes (dealt with in detail in Section 3.3).

The standard enthalpy change for the ionization of water is $+55.83$ kJ mol^{-1}, which means that the reverse reaction, which occurs when acids are neutralized by bases, is exothermic, $i.e.$ $\Delta_r H^\circ = -55.83$ kJ mol^{-1}. The corresponding change in standard Gibbs energy is -79.9 kJ mol^{-1}. The reaction:

$$H^+(aq) + OH^-(aq) \rightarrow H_2O(l) \qquad (1.9)$$

is thermodynamically spontaneous.

Activities of substances are used in strict thermodynamic equations because they properly and accurately represent the deviations from ideality exhibited by those substances when their concentrations (or partial pressures in the gas phase) are not infinitely small.

The sign of the Gibbs energy change for the reaction governs reaction spontaneity; if $\Delta_r G$ is negative the reaction should occur spontaneously (some such reactions are restricted kinetically); if $\Delta_r G$ is positive the reaction is not spontaneous and would not occur under the prevailing conditions.

Equation (1.9) is also one of the most rapid chemical reactions. The second-order rate constant is one of the largest on record, 1.4×10^{11} $dm^3\ mol^{-1}\ s^{-1}$ at 25 °C. The reaction rate is diffusion controlled, *i.e.* the rate depends on the rate of diffusion of the reactants towards each other rather than their chemical characteristics, and there is a reaction every time the reactants meet.

The very negative values of the thermodynamic properties of water given in Table 1.3 (the standard enthalpy of formation, $\Delta_f H°$, and the standard Gibbs energy of formation, $\Delta_f G°$) indicate the considerable thermodynamic stability of the substance compared with the constituent elements in their standard states. The contributions to the value of the standard enthalpy of formation of liquid water may be calculated from data for bond energy terms and the known enthalpy of vaporization (latent heat) of the liquid. The formation of gaseous water from dihydrogen and dioxygen in their standard states may be represented by the equations:

$$H_2(g) \rightarrow 2H(g) \quad 2\Delta_a H°(H,\ g) = 2 \times 218\ kJ\ mol^{-1} \qquad (1.10)$$

$$\tfrac{1}{2}O_2(g) \rightarrow O(g) \quad \Delta_a H°(O,\ g) = 248\ kJ\ mol^{-1} \qquad (1.11)$$

$$2H(g) + O(g) \rightarrow H_2O(g) \quad \Delta_r H° = -(2 \times 463) = -926\ kJ\ mol^{-1} \quad (1.12)$$

The amount $463\ kJ\ mol^{-1}$ represents the enthalpy released when an O$-$H bond is formed; it is the bond enthalpy term for O$-$H single bonds.

The sum of the bond-breaking and bond-making stages gives the result:

$$H_2(g) + \tfrac{1}{2}O_2(g) \rightarrow H_2O(g) \qquad (1.13)$$

for which $\Delta_f H°(H_2O,\ g) = (2 \times 218) + 248 - 926 = -242\ kJ\ mol^{-1}$. The standard enthalpy of vaporization of water is $+44\ kJ\ mol^{-1}$, so the liquid substance is $44\ kJ\ mol^{-1}$ more stable than the gaseous form:

$$H_2O(g) \rightarrow H_2O(l) \qquad (1.14)$$

Note the relatively large difference between the values of $\Delta_v G°$ (H_2O, l) = 8 kJ mol^{-1} and $\Delta_v H°$ (H_2O, l) = 44 kJ mol^{-1}. This is because the $\Delta_v G°$ value includes the $-T\Delta_v S°$ term, and since the evaporation is accompanied by a large increase in entropy the $\Delta_v G°$ value is considerably smaller than the $\Delta_v H°$ value.

and has $\Delta_f H°(H_2O,\ l) = -242 - 44 = -286\ kJ\ mol^{-1}$. The relatively high value for the enthalpy of vaporization arises from the extensive hydrogen bonding in the liquid phase. The thermodynamic stability of liquid water is thus shown to be mainly due to the greater bond strength of the O$-$H bond compared with the strength of the H$-$H bond and half of the strength of the O$=$O bond, and is complemented by the high value of the enthalpy of vaporization of the liquid. The corresponding values for the Gibbs energy quantities are: $\Delta_f G°(H_2O,\ l) = -237\ kJ\ mol^{-1}$, with a contribution of $-8\ kJ\ mol^{-1}$ from the reverse of the Gibbs energy of vaporization, $\Delta_v G°(H_2O,\ l) = +8\ kJ\ mol^{-1}$.

Worked Problem 1.4

Q From the data given in the text, calculate the entropy change when liquid water becomes gaseous under standard conditions at 298 K.

A $\Delta_v G^\circ = \Delta_v H^\circ - T\Delta_v S^\circ$
i.e. $8 = 44 - (298 \times \Delta_v S^\circ)$; therefore $\Delta_v S^\circ = 120.8 \text{ J K}^{-1} \text{ mol}^{-1}$.

Summary of Key Points

1. The terms solvent, solute, solution and solubility were introduced and defined.

2. The structures of the isolated water molecule, ice and liquid water were described.

3. Hydrogen bonding in liquid water was discussed. The significance of the dipolar character of the water molecule was pointed out and its relation to the large value of the permittivity of the bulk liquid.

4. The ionic product or autoprotolysis constant of water, relating to its self-ionization, was discussed.

5. The thermodynamic stabilities of gaseous and liquid water were emphasized.

References

1. J. Barrett, *Chemistry in Your Environment*, Albion Publishing, Chichester, 1994.
2. J. Barrett, *Atomic Structure and Periodicity*, RCS Tutorial Chemistry Text No. 9, The Royal Society of Chemistry, Cambridge, 2002.

Further Reading

F. Franks (ed.), *Water – A Comprehensive Treatise*, Plenum Press, New York, 1973.

Problems

1.1. Derive expressions for (i) molar concentration, (ii) molality and (iii) mole fraction, of a solute whose solubility is given by its mass $\% = 100 m_S/(m_S + m_W)$, where $m_S = $ the mass of solute in the saturated solution and $m_W = $ mass of the water in the solution. Take ρ as the density of the solution in kg m^{-3}.

1.2. The table below contains solubility and density data for the salts Na_2SO_4 and $MgSO_4$. Express their solubilities in terms of molar concentrations, molalities and mole fractions. Calculate the contractions in volume that occur when the solutions are made from the solid salts and the solvent. Comment on the results in terms of the effect of ionic charges. The concentrations have been chosen to be comparable.

Salt	Mass % in solution	Density of solution (kg m^{-3})	Density of solid salt (kg m^{-3})
Na_2SO_4	8	1071.3	2700
$MgSO_4$	10	1103.4	2660

1.3. The melting and boiling temperatures for sodium, water and dioxygen are:

Substance	M.p. ($^\circ$C)	B.p. ($^\circ$C)
Sodium	97.8	883
Water	0	100
Dioxygen	-218.8	-183

What are the standard states for these substances at 25 $^\circ$C?

2

Liquid Water and the Hydration of Ions

This chapter consists of a description of the structure of liquid water and the nature of ions in aqueous solution. The discussion is largely restricted to the interactions between monatomic ions with liquid water in which they become **hydrated** by acquiring a **hydration sphere** or **shell**. Additionally, a few diatomic and polyatomic anions are dealt with, including the important hydroxide ion. The **hydration** of ions derived from the s- and p-block elements of the Periodic Table, and the derivation of values of their **enthalpies** and **entropies of hydration**, are described in considerable detail.

Treatments of the hydration of ions derived from the elements of the d- and f-block elements are contained in Chapters 7 (d-block elements) and 8 (f-block elements).

Aims

By the end of this chapter you should understand:

- That liquid water is hydrogen bonded and has a transient localized structure derived from the more ordered structure of the molecules in ice
- The nature of hydration of cations and anions
- The definition of molar enthalpies and molar entropies of hydration of ions
- The derivation of values of conventional and absolute molar enthalpies and molar entropies of hydration of ions
- The factors influencing the values of molar enthalpies and molar entropies of hydration of ions

2.1   The Structure of Liquid Water

The normal form of ice has an open structure, as described in Chapter 1. The oxygen atoms in the solid are surrounded by an approximately tetrahedral arrangement of four hydrogen atoms, two of which are covalently bonded to the oxygen atom; the other two are hydrogen bonded, but covalently bonded to the oxygen atoms of adjacent water molecules. In liquid water there is some collapse of the crystalline structure, as shown by the increase in density compared to that of ice at 0 °C, but there is major retention of close-range order that persists to varying degrees up to the boiling point. This ensures that the boiling point is anomalously high compared with the other dihydrides of Group 16. There is constant interchange of hydrogen atoms between their oxygen atom partners, so that any particular environment for a water molecule is only transient. The interchange is rapid enough to allow only one proton NMR absorption line. The hydrogen bonding between two molecules of water is illustrated in Figure 2.1.

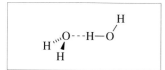

Figure 2.1 Two water molecules participating in hydrogen bonding

The distance between the two oxygen atoms is 276 pm, which is only slightly smaller than twice the van der Waals radius of oxygen, 280 pm. The stability of the hydrogen bond is due to an enhanced dipole–dipole interaction, because of the attractive force between the partial negative charge ($-0.66e$) on the oxygen atom and the partial positive charge on the hydrogen atom ($+0.33e$). The hydrogen nucleus is not efficiently shielded as are other nuclei with their closed shell electronic configurations (*e.g.* the nucleus of Li^+ is shielded by its $1s^2$ configuration), and exerts a major electrostatic contribution to the hydrogen bond stability.

2.2   Ions and their Hydration

When ionic compounds dissolve in water, their constituent positively and negatively charged ions become separated and hydrated in the solution (see equation 1.3). Many covalent compounds also dissolve in water to give solutions containing ions. For example, when hydrogen chloride dissolves in water, the covalent molecule dissociates **heterolytically** (*i.e.* the chlorine atom retains both the bonding electrons to become an anion) to give a solution containing hydrated protons and hydrated chloride ions:

$$HCl(g) \rightarrow H^+(aq) + Cl^-(aq) \tag{2.1}$$

When sulfur(VI) oxide dissolves in water, a solution containing hydrated protons and hydrated hydrogen sulfate(VI) ions [HSO_4^- (aq)] is produced, with some of the latter ions dissociating to give more hydrated protons and hydrated sulfate(VI) ions:

$$SO_3(g) + H_2O \rightarrow H^+(aq) + HSO_4^-(aq) \tag{2.2}$$

$$HSO_4^-(aq) \rightleftharpoons H^+(aq) + SO_4^{2-}(aq) \tag{2.3}$$

In general, ionic compounds act as true solutes in that evaporation of the solvent yields the original compounds. Evaporation of a solution of HCl will give gaseous HCl, but evaporation of the SO_3 solution gives sulfuric(VI) acid, H_2SO_4.

These examples are sufficient to form the basis of the discussion of the nature of cations and anions in aqueous solutions. In aqueous solution, ions are stabilized by their interaction with the solvent; they become **hydrated** and this state is indicated in equations by the **(aq)** symbolism. This is a broad generalization and is amplified by further discussions in the next sections and in the remainder of the book.

2.2.1 The Nature of Ions in Aqueous Solutions

Current ideas of ions in aqueous solutions consist of models that are consistent with electrostatic interactions between the ions and dipolar water molecules and with their effects on the localized structure of liquid water in the immediate vicinity of the ions. Ions in solid crystals have particular and individual sizes that are dependent upon the method used to assign the fractions of the minimum interionic distances to the participating cations and anions. The generally accepted values for ionic radii in crystals are based on the assumption that the ionic radius of the oxide ion is 140 pm. In an aqueous solution the individual ions have an interaction with the solvent that is associated with their ionic radius, but the resulting hydrated ion is difficult to quantify. Geometric considerations limit the number of water molecules that may be associated as immediate neighbours of the hydrated ion. The small number of immediate neighbours around a hydrated ion is known as the **coordination number**, a number that may have values from 4 to 9 for cations, depending on the ion size. Such values are determined by proton NMR spectroscopy. There are separate resonances arising from the bound water molecules and those in the bulk solution.

A simple calculation reveals the limits of the numbers of water molecules that may be associated with an ion in a standard solution. A 1 mol dm^{-3} aqueous solution of sodium chloride has a density of 1038 kg m^{-3} at 25 °C, so 1 dm^3 of such a solution has a mass of 1038 g. One mole of the salt has a mass of 58.44 g, so the water in the litre of solution has a mass of $1038 - 58.44 = 979.56$ g. This amount of water contains $979.56/18.015 = 54.4$ moles of the liquid. The molar ratio of water molecules to ions in the 1 mol dm^{-3} aqueous solution of Na$^+$(aq) and Cl$^-$(aq) ions is therefore $54.4/2 = 27.2$, assuming that the water molecules are shared equally between the cations and anions. This represents the theoretical upper limit of hydration of any one ion in a standard solution of 1 mol dm^{-3} concentration. The limit may be exceeded in more dilute solutions, but that depends upon the operation of forces over a relatively long range. Certainly, in more concentrated solutions, the limits of hydration of ions become more restricted as fewer water molecules are available to share out between the cation and anion assembly.

Some sources of data refer to the state of ions at infinite dilution. In such an extrapolated state the individual ions can exert their full and independent effects upon the solvent molecules.

Worked Problem 2.1

Q The density of a 1 mol dm^{-3} solution of magnesium chloride, $MgCl_2$, at 25 °C is 1066 kg m^{-3}. Calculate the number of water molecules that are available to interact with each ion produced in the solution.

A One dm^3 of a 1 mol dm^{-3} solution of magnesium chloride has a mass of $10^{-3} \times 1066$ kg $= 1066$ g. It contains $24.3 + 70.9 = 95.2$ g of the salt, leaving $1066 - 95.2 = 970.8$ g of water. This amount of water contains $970.8/18.015 = 53.89$ moles. There are three moles of ions in the solution, so each ion has a maximum of $53.89/3 = 18.0$ water molecules that could be in its hydration sphere. It would be expected that the doubly charged Mg^{2+} ion would affect more water molecules than the singly charged chloride ions.

The term **coordinate bond** is synonymous with **dative bond**. Such bonds are formed when one participating partner supplies both electrons to form the single bond. For example, NH_3 forms an adduct with BF_3 in which the non-bonding (lone) pair of the nitrogen atom is used to form a bond to the boron atom, using its otherwise vacant valence orbital: $H_3N{\rightarrow}BF_3$.

Table 2.1 Maximum co-ordination numbers for cations based on valence considerations

Valence orbitals available	Maximum coordination number of cation
s and p	4
s, p and d	9

The water molecules forming the immediate neighbours of a cation may be regarded as forming **coordinate bonds** to the cation, the higher energy non-bonding pair of electrons of the water molecule being used to form the coordinate bond. If this is the case, then the available vacant orbitals of the cation will determine the maximum coordination number. The actual coordination number achieved is the result of the optimization of the electronic and geometric factors. The maximum coordination numbers, dependent upon the valence orbitals available, are given in Table 2.1.

The water molecules that are nearest neighbours of cations form the **primary hydration sphere**. Some indication of the likely content of the primary hydration spheres of cations is given by the numbers of water molecules of crystallization in their solid compounds. For example, the compound $BeSO_4 \cdot 4H_2O$ is better formulated as a **coordination complex** $[Be(H_2O)_4]SO_4$, which indicates that the four water molecules are coordinatively bonded to the Be^{2+} ion. The Be^{2+} ion has an estimated radius of 31 pm and would be expected to form largely covalent bonds, or in aqueous solution to interact strongly with four water molecules to give the more stable **complex ion**, $[Be(H_2O)_4]^{2+}$. The four water molecules are expected to be tetrahedrally arranged around the central Be^{2+} ion, both from a ligand–ligand repulsion viewpoint and also if the acceptor orbitals for the four coordinate bonds are the 2s and three 2p atomic orbitals of Be^{2+}. Solid compounds of lithium such as $LiNO_3 \cdot 3H_2O$ and $LiI \cdot 3H_2O$ seem to imply that there are three water molecules associated with the lithium cations. This is not the case, as the coordination of the lithium cations consists of six water molecules hexagonally disposed around each ion. There are chains of linked lithium cations in which three water molecules are shared as linking groups between two adjacent cations. The octahedral units share trigonal faces. In this case there are insufficient valence orbitals to enable the formation of coordinate bonds, and so the interactions must be regarded as purely electrostatic.

The primary hydration spheres of the cations of the third period, Na^+, Mg^{2+} and Al^{3+}, probably consist of six water molecules, octahedrally arranged and possibly making use of the 3s, 3p and two of the 3d atomic orbitals of the cation. Those of the $+2$ and $+3$ ions of the transition elements are usually six-coordinate octahedrally arranged hexaaqua complexes, $[M(H_2O)_6]^{2+/3+}$, again making use of s, p and d combinations of the atomic orbitals of the central metal ion. Many of the $+2$ and $+3$ oxidation states of the transition elements form crystalline compounds that have at least six water molecules of crystallization per transition metal ion, *e.g.* $FeSO_4 \cdot 7H_2O$. This can be formulated as $[Fe(H_2O)_6]SO_4 \cdot H_2O$, with the seventh water molecule existing outside the coordination sphere of the Fe^{2+} ion and hydrogen bonded to the sulfate ions in the lattice. The primary hydration spheres of the $+3$ ions of the lanthanide f-block elements probably contain nine water molecules. Several of their $+3$ bromate(V) compounds crystallize with nine water molecules of crystallization, *e.g.* $[Sm(H_2O)_9]BrO_3$, in which the nine water molecules are arranged in the form of a tri-capped trigonal prism around the central metal ion, as shown in Figure 2.2.

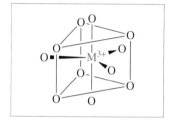

Figure 2.2 A tri-capped trigonal prismatic arrangement of nine water molecules in the primary hydration sphere of a lanthanide +3 ion, showing the positions of the oxygen atoms

There is a conceptual model of hydrated ions that includes the primary hydration shell as discussed above. A **secondary hydration sphere** consists of water molecules that are hydrogen bonded to those in the primary shell and experience some electrostatic attraction from the central ion. This secondary shell merges with the bulk liquid water. A diagram of the model is shown in Figure 2.3. X-ray diffraction measurements and NMR spectroscopy have revealed only two different environments for water molecules in solution of ions. These are associated with the primary hydration shell and water molecules in the bulk solution. Both methods are subject to deficiencies, because of the generally very rapid exchange of water molecules between various positions around ions and in the bulk liquid. Evidence from studies of the electrical conductivities of ions shows that when ions move under the influence of an electrical gradient they "tow" with them as many as 40 water molecules, in dilute solutions.

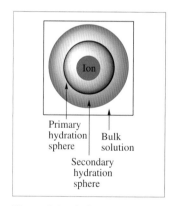

Figure 2.3 A diagram showing the primary and secondary hydration spheres around an ion

The hydration of anions is regarded as being electrostatic with additional hydrogen bonding. The number of water molecules in the primary hydration sphere of an anion depends upon the size, charge and nature of the species. Monatomic anions such as the halide ions are expected to have primary hydration spheres similar to those of monatomic cations. Many aqueous anions consist of a central ion in a relatively high oxidation state surrounded by one or more covalently bonded oxygen atoms, *e.g.* Cl^IO^-, $Cl^{III}O_2^-$, $Cl^VO_3^-$ and $Cl^{VII}O_4^-$. There can be hydrogen bonding between the ligand oxygen atoms and water molecules in the primary hydration sphere; this restricts the coordination numbers but contributes considerably to the stability of the hydrated ion.

The models may be somewhat deficient, but estimates of the enthalpies and entropies of hydrated ions are accurate and are the subject of Section 2.4.

2.2.2 The Hydrated Proton

The hydration of the proton may be regarded as the initial formation of the hydroxonium ion, followed by further interaction with water molecules. The latter process is similar to that experienced by other cations and anions, but they do not participate in full covalent bond formation when they become hydrated.

The hydrated proton, so far in this text, has been written as $H^+(aq)$, but a bare proton might be expected to interact very strongly with a water molecule to give the ion, H_3O^+, that would also be hydrated and written as $H_3O^+(aq)$. The interaction takes the form of an extra covalent bond between a hydrogen atom and the formally positively charged O^+ ion, itself electronically identical with the nitrogen atom. The H_3O^+ ion is isoelectronic with the NH_3 molecule. The ion H_3O^+ is called variously **hydronium, hydroxonium** or **oxonium**.

The ion H_3O^+ is identifiable in crystalline nitric(V) acid monohydrate, $HNO_3 \cdot H_2O$, which is better formulated as $H_3O^+NO_3^-$. The proton makes use of the higher energy non-bonding pair of electrons of the water molecule to form a covalent bond to the oxygen atom. The ion is trigonally pyramidal with HOH bond angles of 112°, not much larger than those of a regular tetrahedron. The bond angle is found to depend upon the hydrogen bonding that occurs in the solid state and is, for instance, 126° in crystalline $H_3O^+ \cdot HSO_4^-$. In solid $HCl \cdot 2H_2O$, better formulated as $[H_5O_2]Cl$, there are dihydrated protons, $H_5O_2^+$ or $(H_2O)_2H^+$, in which the proton is midway between two water molecules, as shown in Figure 2.4.

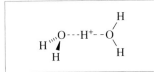

Figure 2.4 A possible structure of the $H_5O_2^+$ ion

Recent work[3,4] has shown that the most stable structure for $H_5O_2^+$ in the gas phase (as an isolated ion) is that shown in Figure 2.5.

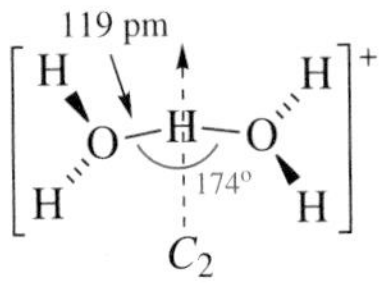

Other hydrated forms of the proton are found in the crystalline state, *e.g.* $H_7O_3^+$ or $(H_2O)_3H^+$ and $H_9O_4^+$ or $(H_2O)_4H^+$, and the largest one is $H_{13}O_6^+$ or $H(H_2O)_6^+$. The protonated oxygen atom in these ions does not take part in hydrogen bonding; all hydrogen bonding to the primarily attached water molecules occurs between the three hydrogen atoms of the H_3O^+ ion. This may be understood in terms of the partial positive charge on the oxygen atom, which reduces the atom's tendency to participate in hydrogen bonding. Further water molecules are hydrogen bonded to the three water molecules already attached to the H_3O^+ ion. In aqueous solution the form $H_3O^+(aq)$ is the accepted way of representing the hydrated proton, but sometimes the shorter form $H^+(aq)$ is used to mean the same species. For example, the autoprotolysis of water is expressed as:

Figure 2.5 The structure of $H_5O_2^+$ in the gas phase showing the C_2 axis

$$H_2O(l) \rightleftharpoons H^+(aq) + OH^-(aq) \qquad (2.4)$$

or:

$$2H_2O(l) \rightleftharpoons H_3O^+(aq) + OH^-(aq) \qquad (2.5)$$

Not so much attention is given to the form of the hydrated hydroxide anion, which is also subject to hydrogen bonded interactions and does exist in some crystals as $H_3O_2^-$ or $(H_2O)OH^-$ and has a symmetrical $[HO–H–OH]^-$ hydrogen bonded structure. Recent work[5] has shown that the most probable form of the hydrated hydroxide ion is $(H_2O)_3OH^-$, in which the hydroxide oxygen atom participates in hydrogen bonds with three water molecules. Exclusively, the hydrated hydroxide ion is formulated as $OH^-(aq)$ in chemical equations.

The central proton is halfway between the two oxygen atoms of the water molecules, and the ion has C_2 symmetry. The hydrogen bonds have lengths of 199 pm and the O–H–O bond angle is 174°. The structure probably represents the transition state in the autoprotolysis of water in which a proton is transferred from one water molecule to a neighbouring one.

2.3 Gibbs Energies and Enthalpies of Formation of Hydrated Ions

This section is devoted to some basic thermodynamics, essential for the understanding of subsequent discussions.

2.3.1 Thermodynamic Conventions

The conventional thermodynamic standard state values of the Gibbs energy of formation and standard enthalpy of formation of elements in their standard states are: $\Delta_f G^\circ = 0$ and $\Delta_f H^\circ = 0$. Conventional values of the standard molar Gibbs energy of formation and standard molar enthalpy of formation of the hydrated proton are: $\Delta_f G^\circ(H^+, aq) = 0$ and $\Delta_f H^\circ(H^+, aq) = 0$. In addition, the standard molar entropy of the hydrated proton is taken as zero: $S^\circ(H^+, aq) = 0$. This convention produces negative standard entropies for some ions.

Throughout this text there are many references to "standard Gibbs energy", "standard enthalpy change" and "standard entropy change". Strictly these should be prefaced by the term "molar" or "partial molar", the latter referring to constituents of mixtures such as solutions of ions. These terms are omitted to save space and the values themselves are always quoted with mol^{-1} in their units.

2.3.2 Standard Values for $\Delta_f G^\circ$ and $\Delta_f H^\circ$ for Some Aqueous Ions

This section contains data for the standard values of the Gibbs energies of formation and enthalpies of formation of some aqueous ions. The manner of their derivation is described briefly.

The **third law of thermodynamics** states that the entropy of a perfect crystalline solid at 0 K is zero. All chemical compounds therefore have standard entropies that are positive. Those associated with ions may be negative, because of the convention of regarding the standard entropy of the hydrated proton as zero, and because they include the effect of the hydration process on the water molecules in their coordination spheres.

Anions

Table 2.2 contains the standard Gibbs energies of formation and the standard enthalpies of formation of a selection of anions at 298.15 K (25 °C).

They refer to the formation of 1 mol dm^{-3} solutions of the anions from their elements in their standard states and are relative to the values for the hydrated proton taken as zero.

Table 2.2 Standard molar Gibbs energies of formation and standard molar enthalpies of formation for some anions at 25 °C (in kJ mol^{-1})

Ion	$\Delta_f G°$	$\Delta_f H°$	Ion	$\Delta_f G°$	$\Delta_f H°$
F$^-$	− 278.8	− 332.6	OH$^-$	− 157.2	− 230.0
Cl$^-$	− 131.2	− 167.2	CN$^-$	+172.4	+150.6
Br$^-$	− 104.0	− 121.6	NO$_3^-$	− 111.3	− 207.4
I$^-$	− 51.6	− 55.2	HCO$_3^-$	− 586.8	− 692.0
ClO$_3^-$	− 8.0	− 104.0	CO$_3^{2-}$	− 527.8	− 677.1
BrO$_3^-$	+18.6	− 67.1	HSO$_4^-$	− 755.9	− 887.3
IO$_3^-$	− 128	− 221.3	SO$_4^{2-}$	− 744.5	− 909.3
ClO$_4^-$	− 8.5	− 129.3	PO$_4^{3-}$	− 1018.7	− 1277.4

The data are derived from thermodynamic information about standard reactions, such as the one for the formation of hydrated protons and hydrated chloride ions from dihydrogen and dichlorine:

$$\tfrac{1}{2}H_2(g) + \tfrac{1}{2}Cl_2(g) \rightarrow H^+(aq) + Cl^-(aq) \tag{2.6}$$

for which $\Delta_r H°$ is − 167.2 kJ mol^{-1}. Because dihydrogen and dichlorine are elements in their standard states and the standard enthalpy of formation of the hydrated proton is taken to be zero, the overall enthalpy of the reaction is identical to the standard enthalpy of formation of the hydrated chloride anion.

When strong bases neutralize strong acids in solutions that have molar concentrations of 1 mol dm^{-3}, the enthalpy of the reaction is observed to be −55.83 kJ mol^{-1}, irrespective of the **counter ions** (*e.g.* the chloride ion derivable from HCl and the sodium ion contained in NaOH) present. For example, when a standard solution (1 mol dm^{-3}) of hydrochloric acid is neutralized by a standard solution (1 mol dm^{-3}) of sodium hydroxide, the change in enthalpy of the reaction is −55.83 kJ mol^{-1}. Because the strong acid HCl and the strong base NaOH are 100% dissociated in aqueous solution, the **neutralization** reaction may be written as:

$$H^+(aq) + Cl^-(aq) + Na^+(aq) + OH^-(aq)$$
$$\rightarrow Na^+(aq) + Cl^-(aq) + H_2O(l) \tag{2.7}$$

The counter ions, Na$^+$(aq) and Cl$^-$(aq), are effectively **spectator ions** in that they do not affect the thermodynamics of the reaction and appear on both sides of equation (2.7). In many instances it is better to dispense with spectator ions in ionic equations, the neutralization reaction reducing to:

$$H^+(aq) + OH^-(aq) \rightarrow H_2O(l) \qquad (2.8)$$

which expresses the chemistry occurring exactly. The value for $\Delta_r H^{\circ}$ for the neutralization reaction is -55.83 kJ mol^{-1} and represents the difference in standard enthalpy of formation of liquid water (-285.83 kJ mol^{-1}) and that of the hydrated hydroxide ion, $\Delta_f H^{\circ}(OH^-, aq)$. The value of the latter quantity is calculated as: $\Delta_f H^{\circ}(OH^-, aq) = -285.83 + 55.83 = -230.0$ kJ mol^{-1}.

Cations

The results for the enthalpies of hydrated hydroxide and chloride anions may then be used to estimate the enthalpies of formation of the cations formed from soluble hydroxides or chlorides. For example, the reaction of sodium with liquid water:

$$Na(s) + H_2O(l) \rightarrow Na^+(aq) + OH^-(aq) + \tfrac{1}{2} H_2(g) \qquad (2.9)$$

has a standard enthalpy change of -184.3 kJ mol^{-1}, which is the sum of the standard enthalpies of formation of the two ions minus the standard enthalpy of formation of liquid water:

$$-184.3 = \Delta_f H^{\circ} (Na^+, aq) + \Delta_f H^{\circ}(OH^-, aq) - \Delta_f H^{\circ}(H_2O, l)$$

which gives a value for $\Delta_f H^{\circ}(Na^+, aq) = -184.3 + 230 - 285.8 = -240.1$ kJ mol^{-1}.

Worked Problem 2.2

Q Given that the standard enthalpy change for dissolving potassium chloride in water is $+17.0$ kJ mol^{-1} and the standard enthalpy of formation of KCl is -436.5 kJ mol^{-1}, use the value for the enthalpy of formation of the aqueous chloride ion given in Table 2.2 to calculate a value for the enthalpy of formation of aqueous potassium ions.

A The enthalpy change for the reaction: $KCl(s) \rightarrow K^+(aq) + Cl^-(aq)$ is $+17.0$ kJ mol^{-1} and is the difference between the enthalpies of formation of the dissolved ions and the solid salt:

$$\Delta_f H^{\circ}(K^+, aq) + \Delta_f H^{\circ}(Cl^-, aq) - \Delta_f H^{\circ}(KCl, s) = +17.0$$
$$\Delta_f H^{\circ}(K^+, aq) = -\Delta_f H^{\circ}(Cl^-, aq) + \Delta_f H^{\circ}(KCl, s) + 17.0$$
$$= 167.2 - 436.5 + 17.0 = -252.3 \text{ kJ mol}^{-1}$$

Table 2.3 contains the standard Gibbs energies of formation and the standard enthalpies of formation of a selection of main group cations at 25 °C. They refer to the formation of 1 mol dm^{-3} solutions of the cations from their elements and are relative to the values for the hydrated proton taken as zero.

No further comments on the values of Table 2.3 are necessary at this stage. The relevance of the data is developed in subsequent sections of the text.

Table 2.3 Standard molar Gibbs energies of formation and standard molar enthalpies of formation of some main group cations at 25 °C (in kJ mol^{-1})

Ion	$\Delta_f G^\circ$	$\Delta_f H^\circ$	Ion	$\Delta_f G^\circ$	$\Delta_f H^\circ$
Li$^+$	−293.3	−278.5	Al^{3+}	−485.0	−531.0
Na$^+$	−261.9	−240.1	Ga^{3+}	−159.0	−211.7
K$^+$	−283.3	−252.4	In^{3+}	−98.0	−105.0
Rb$^+$	−284.0	−251.2	Tl^{3+}	+214.6	+196.6
Cs$^+$	−292.0	−258.2	Tl$^+$	−32.4	+5.4
Be^{2+}	−379.7	−382.8			
Mg^{2+}	−454.8	−466.9	Sn^{2+}	−27.2	−8.8
Ca^{2+}	−553.6	−542.8	Pb^{2+}	−24.4	−1.7
Sr^{2+}	−559.5	−545.8			
Ba^{2+}	−560.8	−537.6			
Ra^{2+}	−561.5	−527.6			

2.4   Standard Molar Enthalpies of Hydration of Ions

The ionic compound caesium chloride, Cs^+Cl^-, dissolves readily in water to give a solution containing the individually hydrated $Cs^+(aq)$ and $Cl^-(aq)$ ions. The thermodynamic parameters for the formation reaction of Cs^+Cl^- and for the reaction of its solution in water are:

	$\Delta_r H^\circ$/kJ mol^{-1}	$\Delta_r G^\circ$/kJ mol^{-1}
$Cs(s) + \frac{1}{2}Cl_2(g) \rightarrow Cs^+Cl^-(s)$	−443.0	−414.5
$Cs^+Cl^-(s) \rightarrow Cs^+(aq) + Cl^-(aq)$	+17.6	−8.7

The reaction for the formation of the hydrated ions is given as the sum of the two reactions above, and the values of $\Delta_r H^\circ$ and $\Delta_r G^\circ$ are derived also by their respective additions of those for the two contributing reactions:

$$Cs(s) + \frac{1}{2} Cl_2(g) \rightarrow Cs^+(aq) + Cl^-(aq) \tag{2.10}$$

$$\Delta_r H^\circ = -443.0 + 17.6 = -425.4 \text{ kJ mol}^{-1}$$

$$\Delta_r G^\circ = -414.5 - 8.7 = -423.2 \text{ kJ mol}^{-1}$$

The enthalpy of formation of the aqueous chloride ion is known to be − 167.2 kJ mol^{-1} (Table 2.2), so that of the aqueous caesium ion may be calculated from the enthalpy of formation of the dissolved $Cs^+(aq)$ and $Cl^-(aq)$ ions as: $\Delta_f H^{\ominus}(Cs^+, aq) = -425.4 - (-167.2) = -258.2$ kJ mol^{-1}.

The enthalpies of formation of aqueous ions may be estimated in the manner described, but they are all dependent on the assumption of the reference zero: that the enthalpy of formation of the hydrated proton is zero. In order to study the effects of the interactions between water and ions, it is helpful to estimate values for the enthalpies of hydration of individual ions, and to compare the results with ionic radii and ionic charges. The **standard molar enthalpy of hydration** of an ion is defined as the enthalpy change occurring when one mole of the gaseous ion at 100 kPa (1 bar) pressure is hydrated and forms a standard 1 mol dm^{-3} aqueous solution, *i.e.* the enthalpy changes for the reactions: $M^{z+}(g) \rightarrow M^{z+}(aq)$ for cations, $X^{z-}(g) \rightarrow X^{z-}(aq)$ for monatomic anions, and $XO_y^{z-}(g) \rightarrow XO_y^{z-}(aq)$ for oxoanions. M represents an atom of an electropositive element, *e.g.* Cs or Ca, and X represents an atom of an electronegative element, *e.g.* Cl or S.

To obtain an estimate of the *combined* enthalpies of hydration of such cations and anions, the thermodynamic cycle shown in Figure 2.6 is helpful. In Figure 2.6 it is assumed that the element X is a halogen and forms a uni-negative ion.

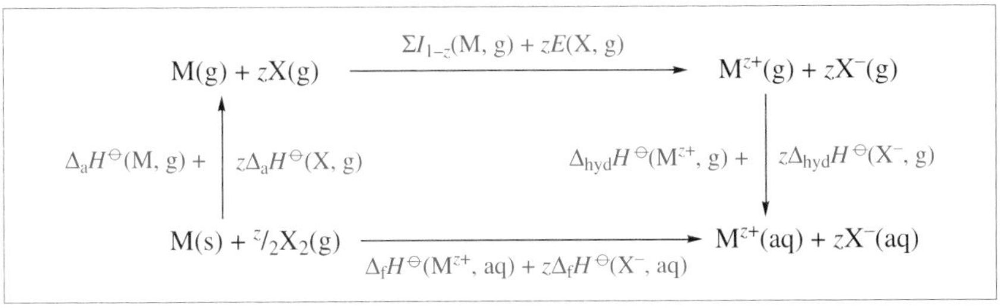

The definition of enthalpy of hydration given in the text is generally used, but the process of hydration consists of two main changes: (i) the compression of the gaseous ion from its **molar volume**, $V_m^{\ominus}$ of 24.79 dm^3 (2.479×10^{-2} m^3 mol^{-1}) to 1 dm^3, followed by (ii) the addition of the water and its interaction with the gaseous ions. $V_m^{\ominus} = RT/p = 24.79$ dm^3 at a temperature of $T = 298.15$ K at a pressure $p = 100$ kPa. A third factor, electrostatic interaction between the charged ions, is not specifically incorporated. The repulsions between like-charged ions would be considerable, but these are offset by the attractions between oppositely charged ions. The latter are equally divided between the two individual hydration enthalpies.

Figure 2.6 A thermochemical cycle for the formation of hydrated ions from their elements

The cycle allows the overall enthalpy of formation of the aqueous solution of cations and anions to be sub-divided into stages whose enthalpy changes are known except for the two enthalpies of hydration, allowing their sum to be estimated. The equation to be solved is:

$$\Delta_f H^{\ominus}(M^{z+}, aq) + z\Delta_f H^{\ominus}(X^-, aq)$$
$$= \Delta_a H^{\ominus}(M, g) + z\Delta_a H^{\ominus}(X, g) + \Sigma I_{1 \rightarrow z}(M, g) + zE(X, g)$$
$$+ \Delta_{hyd} H^{\ominus}(M^{z+}, g) + z\Delta_{hyd} H^{\ominus}(X^-, g) \quad (2.11)$$

which may be rearranged to give the sum of the two enthalpies of hydration:

$$\Delta_{hyd} H^{\ominus}(M^{z+}, g) + z\Delta_{hyd} H^{\ominus}(X^-, g)$$
$$= \Delta_f H^{\ominus}(M^{z+}, aq) + z\Delta_f H^{\ominus}(X^-, aq) - \Delta_a H^{\ominus}(M, g)$$
$$- z\Delta_a H^{\ominus}(X_2, g) - \Sigma I_{1 \rightarrow z}(M, g) - zE(X, g) \quad (2.12)$$

Thermochemical cycles or Born–Haber cycles are extremely useful in the interpretation of chemical changes. They are based on the First Law of Thermodynamics, the Law of Conservation of Energy.

The term $\Delta_a H^{\ominus}(X, g)$ represents the standard enthalpy of formation of one mole of the gaseous atom X(g) from the standard state of the element.

Internal energies and enthalpy changes. Atomic and molecular energy quantities, such as ionization energies, are given in data books as values of **internal energy**, U, or changes in internal energy, ΔU, which are values at **constant volume** and at **0 K**. A change in enthalpy, ΔH, is related to the change in internal energy by the equation: $\Delta H = \Delta U + p\Delta V = \Delta U + \Delta nRT$ (p is pressure, V is volume and Δn is the change in the number of moles of particles). In the case of ionization energies in which an atom is converted into a positive ion plus the liberated electron, $\Delta n = +1$. This means that the corresponding change in enthalpy, ΔH, is given by $\Delta U + RT$, and at the standard temperature of 298 K a further $\frac{3}{2}RT$ is added to allow for the extra enthalpy possessed by the products of ionization at the temperature, T ($\frac{1}{2}RT$ for each of the extra degrees of translational freedom). To convert an ionization energy to an ionization enthalpy, an amount $\frac{5}{2}RT$ (6.2 kJ mol^{-1} at 298.15 K, 25 °C) should be added. Thus the ionization energy of the hydrogen atom, 1312 kJ mol^{-1}, is converted to the enthalpy of ionization by adding 6.2 kJ mol^{-1} to give 1318.2 kJ mol^{-1}. The difference is less than 0.5%, and is normally not of any great significance. In calculations in which atomic properties are mixed with true enthalpy changes, the difference should be acknowledged. In thermochemical cycles referring to overall reactions, the change(s) of ionization energies to enthalpy changes are cancelled out when the electron attachment energy terms are also changed to enthalpies.

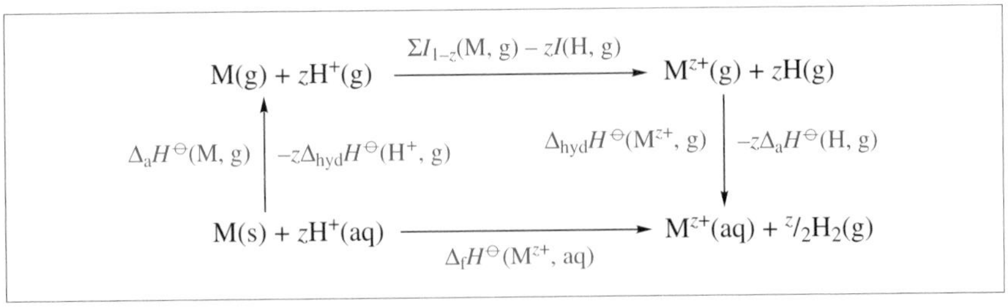

Figure 2.7 A thermochemical cycle for the formation of the hydrated M^{z+} ion

The I terms are the successive ionization energies of the element M, and the E term is the electron attachment energy of the element X.

This treatment illustrates the inherent difficulty of the problem. Any cycle of this type will allow the calculation of *sums* of enthalpies of hydration of cations and anions, but it will not allow the estimation of the separate quantities. There are two ways of dealing with the matter. One is to use the conventional reference zero: that the enthalpy of hydration of the proton is zero, *i.e.* $\Delta_{hyd}H^{\circ}(H^{+}, g) = 0$. The second approach is to estimate an absolute value for $\Delta_{hyd}H^{\circ}(H^{+}, g)$ that may be used to estimate the absolute values for enthalpies of hydration of any other ions. Both are exemplified in the text, but only the second is of general use in the study of the hydration of ions and the discussion of the factors that determine the values of enthalpies of hydration of individual ions.

2.4.1 Conventional Enthalpies of Hydration of Ions

Conventional enthalpies of hydration of ions are those obtained by the first method, *i.e.* those using the *convention* that the enthalpy of hydration of the proton is *zero*.

Cations

Using the same symbolism as the previous sub-section, Figure 2.7 shows a thermodynamic cycle for the formation of the cation of element M by treating the element in its standard state with an acid solution containing hydrated protons and producing the hydrated cation and an equivalent amount of dihydrogen:

$$M(s) + zH^{+}(aq) \rightarrow M^{z+}(aq) + \tfrac{z}{2}H_2(g) \qquad (2.13)$$

The overall change of enthalpy represents the enthalpy of formation of the hydrated cation, M^{z+}(aq), and has a value as given in Tables 2.2 and 2.3 for particular cases. The value consists of contributions from the enthalpy of atomization of element M and the appropriate sum of its ionization enthalpies, and the enthalpy of hydration of the gaseous ion.

It also includes the enthalpy of ionization of the hydrogen atom (equal to, but opposite in sign to, the electron attachment enthalpy of the gaseous proton), the enthalpy of atomization of dihydrogen and the enthalpy of hydration of the proton. The enthalpy of formation of the cation is estimated by use of the equation:

$$\Delta_f H^\circ(M^{z+}, aq) = \Delta_a H^\circ(M, g) - z\Delta_{hyd}H^\circ(H^+, g) + \Sigma_{1\to z}(M, g)$$
$$+ 6.2z - z(I(H, g) + 6.2) + \Delta_{hyd}H^\circ(M^{z+}, g) \quad (2.14)$$
$$- z\Delta_a H^\circ(H, g)$$

The equation may be rearranged to give:

$$\Delta_{hyd}H^\circ(M^{z+}, g) - z\Delta_{hyd}H^\circ(H^+, g) = \Delta_f H^\circ(M^{z+}, aq)$$
$$- \Delta_a H^\circ(M, g) - \Sigma I_{1\to z}(M, g) + zI(H, g) + z\Delta_a H^\circ(H, g) \quad (2.15)$$

Since the second term on the left-hand side of equation (2.15) is conventionally set to zero, the equation may be used to estimate conventional values of enthalpies of hydration for a series of ions. The left-hand side of the equation represents the conventional value of the enthalpy of hydration of the M^{z+} ion:

$$\Delta_{hyd}H^\circ(M^{z+}, g)^{conv} = \Delta_{hyd}H^\circ(M^{z+}, g) - z\Delta_{hyd}H^\circ(H^+, g) \quad (2.16)$$

The terms on the right-hand side of equation (2.15) are known with considerable accuracy. The conventional values for the enthalpies of hydration of the Group 1 cations are estimated from the data given in Table 2.4.

This is an example of the changes of ionization energies to enthalpy changes cancelling out with the equivalent changes to the electron attachment energies of the participating protons in the reaction. The two 6.2z terms cancel out and are omitted from the rearranged equation. [5/2RT = 6.2 kJ mol⁻¹ at 298.15 K]

Table 2.4 Estimated conventional molar enthalpies of hydration for the cations of Group 1 elements (kJ mol⁻¹); I(H, g) = 1312 kJ mol⁻¹, $\Delta_a H^\circ$(H, g) = 218 kJ mol⁻¹

Element	$\Delta_f H^\circ(M^+, aq)$	$\Delta_a H^\circ(M, g)$	$I_1(M, g)$	$\Delta_{hyd}H^\circ(M^+, g)^{conv}$
Li	−278.5	+159	520	+572.5
Na	−240.1	+108	496	+685.9
K	−252.4	+89	419	+769.6
Rb	−251.2	+81	403	+794.8
Cs	−258.3	+77	376	+818.7

The estimated conventional values for the enthalpies of hydration of the Group 1 cations are very positive and give the wrong impression about the interaction of the gaseous ions with liquid water. This is because of the convention for the hydrated proton. The absolute values for the enthalpies of hydration for gaseous cations are given by the equation (rearranged equation 2.16):

$$\Delta_{hyd}H^\circ(M^{z+}, g) = \Delta_{hyd}H^\circ(M^{z+}, g)^{conv} + z\Delta_{hyd}H^\circ(H^+, g) \quad (2.17)$$

when an absolute value for the enthalpy of hydration of the proton is used instead of the conventional zero value. This is likely to be considerably negative, and when included in equation (2.17) would make the absolute values for the enthalpies of the Group 1 cations negative quantities. This is intuitively correct, as it would be expected that an interaction between a positive ion and dipolar water molecules would lead to the release of energy. The hydration process would be expected to be *exothermic*.

Anions

The conventional enthalpies of hydration of the uni-negatively charged halide ions may be estimated using the thermodynamic cycle shown in Figure 2.8.

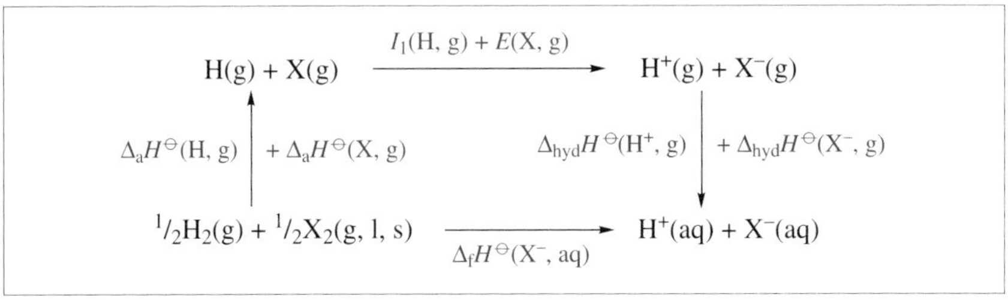

Figure 2.8 A thermochemical cycle for the formation of the hydrated X^- ion

The reactions considered here are the formations of the hydrated halide ions when the halogen molecules react with dihydrogen to give the hydrated proton and the hydrated halide ions.

$$\tfrac{1}{2}H_2(g) + \tfrac{1}{2}X_2(g, 1 \text{ or } s) \rightarrow H^+(aq) + X^-(aq) \qquad (2.18)$$

The overall change of enthalpy represents the enthalpy of formation of the hydrated anion, $X^-(aq)$, and has a value as given in Table 2.2 for particular cases. The value consists of contributions from enthalpy of atomization of element M and its electron attachment energy, and the enthalpy of hydration of the gaseous ion. It also includes the enthalpy of ionization of the hydrogen atom, the enthalpy of atomization of dihydrogen and the enthalpy of hydration of the proton. The enthalpy of formation of the anion is estimated by the equation:

$$\begin{aligned}
\Delta_f H^{\oplus}(X^-, aq) = {} & \Delta_a H^{\oplus}(H, g) + \Delta_a H^{\oplus}(X, g) + (I(H, g) + 6.2) \\
& + (E(X, g) - 6.2) + \Delta_{hyd} H^{\oplus}(H^+, g) \\
& + \Delta_{hyd} H^{\oplus}(X^-, g)
\end{aligned} \qquad (2.19)$$

The equation may be rearranged (cancelling out the 6.2 terms) to give:

$$\begin{aligned}
\Delta_{hyd} H^{\oplus}(X^-, g) + \Delta_{hyd} H^{\oplus}(H^+, g) = {} & \Delta_f H^{\oplus}(X^-, aq) - \Delta_a H^{\oplus}(H, g) \\
& - \Delta_a H^{\oplus}(X, g) - I(H, g) \\
& - E(X, g)
\end{aligned} \qquad (2.20)$$

The term $\Delta_a H^{\oplus}(H, g)$ represents the standard enthalpy of formation of one mole of the gaseous atom H(g) from the standard state of the element [$H_2(g)$].

Since the second term on the left-hand side of equation (2.20) is conventionally set to zero, the equation may be used to estimate conventional values of enthalpies of hydration for a series of halide ions. The left-hand side of the equation (2.20) represents the conventional value of the enthalpy of hydration of the X^- ion:

$$\Delta_{hyd}H^{\circ}(X^-, g)^{conv} = \Delta_{hyd}H^{\circ}(X^-, g) + \Delta_{hyd}H^{\circ}(H^+, g) \qquad (2.21)$$

The terms on the right-hand side of equation (2.20) are known with considerable accuracy. The conventional values for the enthalpies of hydration of the halide anions are estimated from the data given in Table 2.5.

Table 2.5   Estimated conventional molar enthalpies of hydration for the uni-negative anions of the Group 17 elements (kJ mol^{-1}); $I(H, g) = +1312$ kJ mol^{-1}, $\Delta_a H^{\circ}(H, g) = +218$ kJ mol^{-1}

Element	$\Delta_f H^{\circ}(X^-, aq)$	$\Delta_a H^{\circ}(X, g)$	$E(X, g)$	$\Delta_{hyd}H^{\circ}(X^-, g)^{conv}$
$F_2(g)$	−335.4	+79.4	−328	−1616.8
$Cl_2(g)$	−167.2	+121	−349	−1469.2
$Br_2(l)$	−121.4	+112	−325	−1438.4
$I_2(s)$	−56.8	+107	−295	−1398.8

The estimated conventional values for the enthalpies of hydration of the Group 17 halide anions are extremely negative and, as was the case with the Group 1 cations, the wrong impression about the interaction of the gaseous ions with liquid water is given. Intuitively, it might be expected that similarly charged ions, positive or negative, would interact electrostatically with water to roughly the same extent, and that their absolute values of enthalpies of hydration would be of the same order. The apparent discrepancies occur because of the convention regarding the hydrated proton. The absolute values for the enthalpies of hydration for gaseous anions are given by the equation (rearranged equation 2.21):

$$\Delta_{hyd}H^{\circ}(X^-, g) = \Delta_{hyd}H^{\circ}(X^-, g)^{conv} - \Delta_{hyd}H^{\circ}(H^+, g) \qquad (2.22)$$

When an absolute value for the enthalpy of hydration of the proton is used in equation (2.22) it would have the effect of making the absolute enthalpies of hydration of the halide ions considerably less negative than the conventional values.

Pay particular notice to the signs of terms in thermodynamic cycle equations. The electron attachment energies of the halogens (and the great majority of all the elements) are quoted with minus signs. This is because of the definition of the term as "the energy change when a mole of an atom of a gaseous element accepts a mole of electrons to give a mole of gaseous negative ions". A term $+E$ in an equation represents a negative amount of energy. Note also that the two 6.2 kJ mol^{-1} ($=\frac{5}{2}RT$) terms cancel out. [There are a few exceptional cases of elements that have positive electron attachment energies, *e.g.* calcium has a value $E=+186$ kJ mol^{-1}; Ca$^-$(g) is that amount less stable than Ca(g).]

2.5   The Absolute Value for the Standard Molar Enthalpy of Hydration of the Proton

Plots of the conventional values for the enthalpies of hydration of the Group 1 cations and the Group 17 anions against the ionic radii of the ions are shown in Figure 2.9, using the values from Tables 2.4 and 2.5.

Figure 2.9 Plots with trend lines of the conventional enthalpies of hydration of the Group 1 cations and Group 17 anions against their ionic radii; also included are the values of the conventional enthalpies of hydration of the Group 1 cations minus 1110 kJ mol^{-1} and the conventional enthalpies of hydration of the Group 17 anions plus 1110 kJ mol^{-1}

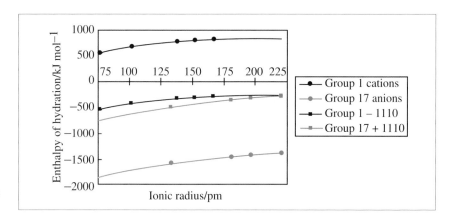

Trend lines are shown which emphasize that the values for $\Delta_{hyd}H^{\circ}(M^{z+}, g)^{conv}$ and $\Delta_{hyd}H^{\circ}(X^{-}, g)^{conv}$ vary with ionic radius. This is to be expected from simple electrostatic arguments. The smaller ions would be expected to have greater interactions with water than the larger ones.

From the values for the conventional enthalpies of hydration of the Group 1 cations (Table 2.4) and those of the halide ions (Table 2.5), it is clear that they differ enormously. This leads to the quest for absolute values, which can be compared on a more equal basis. Equations (2.17) and (2.22) connect the conventional and absolute values of cations and anions respectively, and an approximate value for $\Delta_{hyd}H^{\circ}(H^{+}, g)$ may be obtained by comparing the conventional values of the enthalpies of hydration of the Group 1 cations and the Group 17 halide anions by using equation (2.16) modified for singly charged cations:

$$\Delta_{hyd}H^{\circ}(M^{+}, g)^{conv} = \Delta_{hyd}H^{\circ}(M^{+}, g) - \Delta_{hyd}H^{\circ}(H^{+}, g) \qquad (2.23)$$

and equation (2.21). The difference between $\Delta_{hyd}H^{\circ}(M^{+}, g)^{conv}$ and $\Delta_{hyd}H^{\circ}(X^{-}, g)^{conv}$ [*i.e.* the difference between equation (2.23) and equation (2.21)] is given by the equation:

$$\Delta_{\text{hyd}}H^{\circ}(\text{M}^{+},\text{g})^{\text{conv}} - \Delta_{\text{hyd}}H^{\circ}(\text{X}^{-},\text{g})^{\text{conv}}$$
$$= \Delta_{\text{hyd}}H^{\circ}(\text{M}^{+},\text{g}) - \Delta_{\text{hyd}}H^{\circ}(\text{X}^{-},\text{g}) - 2\Delta_{\text{hyd}}H^{\circ}(\text{H}^{+},\text{g}) \qquad (2.24)$$

This would allow the estimation of the value of $\Delta_{\text{hyd}}H^{\circ}(\text{H}^{+},\text{g})$ *only* if the terms representing the absolute values for the enthalpies of hydration of the two ions cancel out, so that:

$$\Delta_{\text{hyd}}H^{\circ}(\text{M}^{+},\text{g})^{\text{conv}} - \Delta_{\text{hyd}}H^{\circ}(\text{X}^{-},\text{g})^{\text{conv}} = -2\Delta_{\text{hyd}}H^{\circ}(\text{H}^{+},\text{g}) \qquad (2.25)$$

This condition is impossible to realize with existing ions, but an empirical approach yields dividends.

The mean value of the conventional enthalpies of hydration of the Group 1 cations is $+728.3$ kJ mol^{-1} and that of the conventional enthalpies of hydration of the Group 17 anions is -1480.8 kJ mol^{-1}. If it is assumed that the taking of means allows for the variations of the enthalpy terms with ion size and arranges for $\Delta_{\text{hyd}}H^{\circ}(\text{M}^{z+},\text{g})$ to be equal to $\Delta_{\text{hyd}}H^{\circ}(\text{X}^{-},\text{g})$, the difference between the two mean values may be equated to twice the value for the absolute enthalpy of hydration of the proton:

$$728.3 - (-1480.8) = -2\Delta_{\text{hyd}}H^{\circ}(\text{H}^{+},\text{g})$$

giving (rounded off to 4 figures) $\Delta_{\text{hyd}}H^{\circ}(\text{H}^{+},\text{g}) = -1105$ kJ mol^{-1}.

The mean ionic radii of the Group 1 cations and the Group 17 anions are 127 pm and 183 pm, respectively, a difference of 56 pm. Figure 2.10 shows plots similar to those in Figure 2.9, but the values of $\Delta_{\text{hyd}}H^{\circ}(\text{M}^{+},\text{g})$ are plotted at points where the radii of the Group 1 cations are 50 pm greater than their actual ionic radii. Also, the values of $\Delta_{\text{hyd}}H^{\circ}(\text{M}^{z+},\text{g})^{\text{conv}} - 1110$ and $\Delta_{\text{hyd}}H^{\circ}(\text{X}^{-},\text{g})^{\text{conv}} + 1110$ are plotted and trend lines are drawn through the two sets of points. The "shifted" metal ion radii of $(r_i + 50)$ pm and the value of 1110 kJ mol^{-1} for $\Delta_{\text{hyd}}H^{\circ}(\text{H}^{+},\text{g})$ were optimized by the author, using a spreadsheet to arrange that the trend lines for the cation and anion values were as coincident as possible.

Ionic radii and crystal radii. Ionic radii are used exclusively in this text. They are the values of ionic radii which are based on the oxide ion, O^{2-}, having the value 140 pm and are generally additive in that they reproduce minimum interionic distances in solid-state compounds. Crystal radii also apply to solid compounds but have a different basis, dependent upon the assumption that the crystal radii of K^+ and Cl^- are equal. Crystal radii are converted into ionic radii by adding 14 pm to cation radii and subtracting 14 pm from the anion radii.

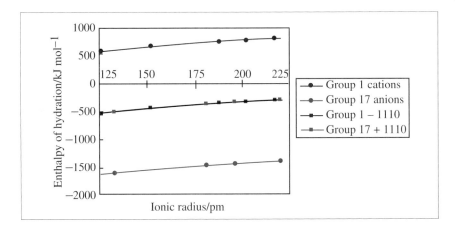

Figure 2.10 Plots and trend lines of the conventional molar enthalpies of hydration of the Group 1 cations and Group 17 anions against their ionic radii; the radii of the Group 1 cations have been increased by 50 pm; also included are the values of the conventional enthalpies of hydration of the Group 1 cations minus 1110 kJ mol^{-1} and the conventional enthalpies of hydration of the Group 17 anions plus 1110 kJ mol^{-1}

The estimation of the value of $\Delta_{hyd}H^{\ominus}(H^+, g)$ has a long and complicated history. The earliest value of -1088 kJ mol^{-1} was obtained by K. Fajans (of Fajans' Rules) in 1919! A very detailed calculation by Halliwell and Nyberg[3] in 1963 gave a value of -1091 kJ mol^{-1}. The most negative value reported is -1250 kJ mol^{-1}. Bockris and Reddy[4] suggest that the "best" value is the one adopted in this text: -1110 kJ mol^{-1}.

Variations in the values of the two parameters used in the spreadsheet operation – the addition to the cation radii and the value of the absolute enthalpy of hydration of the proton – produce a very shallow minimum approximating to the 50 pm addition to the cation radii and the value of $\Delta_{hyd}H^{\ominus}(H^+, g) = -1110$ kJ mol^{-1}. This means that there are considerable uncertainties in both parameters, possibly as much as ± 20 kJ mol^{-1} in the values of the enthalpies of hydration and ± 10 pm in the cation radius shift.

That the trend lines coincide almost completely is some justification of the empirical approach.

Absolute values for the enthalpies of hydration of ions given in the text are calculated using the value $\Delta_{hyd}H^{\ominus}(H^+, g) = -1110$ kJ mol^{-1}.

2.6   Absolute Values of the Standard Molar Enthalpies of Hydration of Ions

By the use of equations (2.17) and (2.22), together with the value of -1110 kJ mol^{-1} for the absolute enthalpy of hydration of the proton, the absolute enthalpies of hydration for the Group 1 cations and Group 17 anions may be estimated. The values are given in Tables 2.6 and 2.7.

Table 2.6   Ionic radii and absolute standard molar enthalpies of hydration for the Group 1 cations

Ion	Ionic radius/pm	$\Delta_{hyd}H^{\ominus}(M^+, g)$/kJ mol^{-1}
Li$^+$	76	$572.5 - 1110 = -538$
Na$^+$	102	$685.9 - 1110 = -424$
K$^+$	138	$769.6 - 1110 = -340$
Rb$^+$	152	$794.8 - 1110 = -315$
Cs$^+$	167	$818.7 - 1110 = -291$

Table 2.7   Ionic radii and absolute standard molar enthalpies of hydration for the Group 17 anions

Ion	Ionic radius/pm	$\Delta_{hyd}H^{\ominus}(X^-, g)$/kJ mol^{-1}
F$^-$	133	$-1614 - (-1110) = -504$
Cl$^-$	181	$-1469.2 - (-1110) = -359$
Br$^-$	196	$-1438.4 - (-1110) = -328$
I$^-$	220	$-1397.2 - (-1110) = -287$

Worked Problem 2.3

Q   Using -1110 kJ mol^{-1} for the absolute value of the enthalpy of hydration of the proton, calculate values for the absolute values of the enthalpies of hydration of the Rb$^+$ and Br$^-$ ions.

A   The absolute value of the enthalpy of hydration of the Rb$^+$ ion is given by the equation (derived from equation (2.17)):

$$\Delta_{hyd}H^{\ominus}(Rb^+, g) = \Delta_{hyd}H^{\ominus}(Rb^+, g)^{conv} + \Delta_{hyd}H^{\ominus}(H^+, g) \quad (2.26)$$

and putting in the value for $\Delta_{hyd}H^{\ominus}(Rb^+, g)^{conv}$ from Table 2.4 gives:

$$\Delta_{hyd}H^{\ominus}(Rb^+, g) = 794.8 - 1110.0 = -315\,kJ\,mol^{-1}.$$

The absolute value of the enthalpy of hydration of the Br^- ion is given by the equation (derived from equation 2.22):

$$\Delta_{hyd}H^{\ominus}(Br^-, g) = \Delta_{hyd}H^{\ominus}(Br^-, g)^{conv} - \Delta_{hyd}H^{\ominus}(H^+, g) \qquad (2.27)$$

and putting in the value for $\Delta_{hyd}H^{\ominus}(Br^-, g)^{conv}$ from Table 2.5 gives:

$$\Delta_{hyd}H^{\ominus}(Br^-, g) = -1438.4 + 1110.0 = -328\,kJ\,mol^{-1}$$

The results for the enthalpies of hydration of the Group 1 cations and the Group 17 anions show that their values become less negative as the ionic radii of the ions increase.

The process of hydration of an ion refers to the conversion of one mole of the gaseous ions under standard conditions at a pressure of 1 bar to the hydrated ions at a molar concentration of $1\,mol\,dm^{-3}$. The process may be divided into two parts. These are the compression of the one mole of gaseous ions into a volume of $1\,dm^3$ followed by the interaction of the ions with water to produce the hydrated ions. Assuming ideal gas behaviour, the compression of one mole of a gas at standard pressure and at 298.15 K into a volume of $1\,dm^3$ requires the expenditure of enthalpy given by $RT\ln(24.79/1.0) = +7.96\,kJ\,mol^{-1}$. The quoted values of ionic hydration enthalpies include a contribution from the compression of the gaseous ions and the enthalpy changes associated with the hydration process are given by the equation:

$$\Delta_{hyd}H^{\ominus}(\text{hydration only}) = \Delta_{hyd}H^{\ominus} - 7.96 \qquad (2.28)$$

The discrepancy introduced by ignoring the compression term is only slight and represents no more than a 2% difference from the absolute values usually used. It is normally ignored in view of the uncertainties associated with the problem of dividing the conventional enthalpies of hydration of cation–anion pairs into individual values.

The conventional enthalpies of hydration of ions are fairly accurate, but the absolute values – which are dependent upon the assumptions and methods used to estimate the value of the absolute enthalpy of hydration of the proton – are subject to greater uncertainties. The more negative the value of the enthalpy of hydration of the proton used, the more negative are the values of the enthalpies of hydration of cations and the less negative are the enthalpies of hydration of anions.

2.6.1 The Born Equations for the Hydration of Ions

Max Born introduced a theoretical approach to the estimation of absolute enthalpies of hydration of ions based on electrostatic theory. Equation (1.2) is fundamental to electrostatic theory. It may be altered to:

$$E_p = q_1\phi \qquad (2.29)$$

Max Born was awarded the 1954 Nobel Prize for Physics for his work on quantum mechanics. He did much work on the ionic state in solids and in aqueous solutions, resulting in the current-day usage of the Born–Landé equation for the calculation of lattice enthalpies (this was modified by Mayer, yielding the sometimes used Born–Mayer equation for the same purpose), and in the Born equations for the calculations of Gibbs energies of hydration and the enthalpies and entropies of hydration of ions described in this section.

where the potential energy of a charge q_1 in the presence of another charge q_2 may be expressed in terms of the Coulomb potential, ϕ:

$$\phi = \frac{q_2}{4\pi\varepsilon_0 r} \tag{2.30}$$

The units of ϕ are J C^{-1} or V (volts). When a potential is multiplied by a charge, as in equation (2.29), the result has units of joules, 1 J = 1 V C. The Born model of an ion is a sphere of charge q and radius r_i in a medium of permittivity ε. The potential at the surface of the sphere is:

$$\phi = \frac{q}{4\pi\varepsilon r_i} \tag{2.31}$$

To charge up such a sphere from 0 to q is calculated as the work, w, expressed by the integral of $\phi \mathrm{d}q$:

$$w = \int_0^q \phi \mathrm{d}q = \frac{1}{4\pi\varepsilon r_i}\int_0^q q\mathrm{d}q = \frac{q^2}{8\pi\varepsilon r_i} \tag{2.32}$$

The work needed to charge the same sphere to the charge $q = ze$ in a vacuum ($\varepsilon = \varepsilon_0$) is:

$$w_{\text{vac}} = \frac{z^2 e^2}{8\pi\varepsilon_0 r_i} \tag{2.33}$$

In a medium with a relative permittivity of ε_r ($\varepsilon = \varepsilon_0\varepsilon_r$), the work needed to charge the sphere to the same extent would be:

$$w_{\text{med}} = \frac{z^2 e^2}{8\pi\varepsilon_0\varepsilon_r r_i} \tag{2.34}$$

The work needed to transfer the ion from a vacuum to the medium is given by:

$$w_{\text{med}} - w_{\text{vac}} = \frac{z^2 e^2}{8\pi\varepsilon_0\varepsilon_r r_i} - \frac{z^2 e^2}{8\pi\varepsilon_0 r_i} \tag{2.35}$$

Multiplying the work of transfer by the Avogadro constant gives the molar change in Gibbs energy of hydration when ε_r is that of liquid water (78.54 at 25 °C). Equation (2.35), incorporating the Avogadro constant, N_A, and the value for the relative permittivity of water, may be rearranged to give:

$$\Delta_{\text{hyd}}G^{\circ} = -\frac{z^2 e^2 N_A}{8\pi\varepsilon_0 r_i}\left(1 - \frac{1}{\varepsilon_{\text{water}}}\right) \tag{2.36}$$

Putting the values of the constants into equation (2.36) gives:

$$\Delta_{\text{hyd}}G^{\circ} = -6.86\times 10^4 \times \frac{z^2}{r_i}\ \text{kJ mol}^{-1} \tag{2.37}$$

when r_i has units of picometres.

Assuming that only ε_{water} varies with temperature, and making use of the relationship:

$$\Delta S = -\frac{\partial(\Delta G)}{\partial T} \qquad (2.38)$$

gives a value for the entropy change accompanying ion hydration as

$$\Delta_{hyd}S^{\circ} = \frac{z^2 e^2 N_A}{8\pi\varepsilon_0\varepsilon_{water}r_i}\frac{\partial(\ln \varepsilon_{water})}{\partial T} \qquad (2.39)$$

The temperature coefficient of the natural logarithm of the relative permittivity of water is –0.0046, and insertion into equation (2.39) gives:

$$\Delta_{hyd}S^{\circ} = -\frac{4069z^2}{r_i} \text{ J K}^{-1}\text{mol}^{-1} \qquad (2.40)$$

when r_i has units of picometres. The negative sign of the entropy change when gaseous ions are hydrated is consistent with the loss of translational freedom as hydration of the ions causes restrictions to the water molecules that participate in the process.

Making use of the relationship:

$$\Delta H^{\circ} = \Delta G^{\circ} + T\Delta S^{\circ} \qquad (2.41)$$

the value of the enthalpy of hydration for an ion with charge z and radius r_i (in picometres) is given by:

$$\Delta_{hyd}H^{\circ} = -\frac{z^2}{r_i}\times(6.86\times10^4 + 4.068T)\text{ kJ mol}^{-1} \qquad (2.42)$$

At 298.15 K the values of enthalpies of hydration are given by:

$$\Delta_{hyd}H^{\circ} = -\frac{z^2}{r_i}\times(6.98\times10^4)\text{ kJ mol}^{-1} \qquad (2.43)$$

The Born equations (2.37) and (2.43) are widely used, but suffer from lack of accurate information about the real sizes of ions in aqueous solutions. In the derivations used above, ionic radii have been used, but these give numerical answers that are exaggerated, in particular for those of cations. The values of the absolute enthalpies of hydration of the ions of Table 2.3 are given in Table 2.8, based on the conventional values with that of the proton taken to be -1110 kJ mol^{-1}.

The values for the enthalpies of hydration of ions may be forced to conform to the expectations of the Born equation (2.43) by estimating a suitable value for each ionic radius or by estimating an apparent ionic charge.

Note that the ratio of the Born entropy of hydration to the Born enthalpy of hydration is extremely small. This is not justified by the treatment of entropies of hydration of ions given in Section 2.7.2, but is indicative of the relatively secondary importance of the hydration entropy in determining the Gibbs energy of hydration. The Born equations assume that the water solvent is a continuous medium and give no credence to its unique properties.

Table 2.8 Standard absolute molar enthalpies of hydration of some main group cations at 25 °C (in kJ mol^{-1})

Ion	$\Delta_{hyd}H^{\circ}$
Li$^+$	−538
Na$^+$	−424
K$^+$	−340
Rb$^+$	−315
Cs$^+$	−291
Be^{2+}	−2524
Mg^{2+}	−1963
Ca^{2+}	−1616
Sr^{2+}	−1483
Ba^{2+}	−1346
Ra^{2+}	−1335
Al^{3+}	−4741
Ga^{3+}	−4745
In^{3+}	−4171
Tl^{3+}	−4163
Tl$^+$	−346
Sn^{2+}	−1587
Pb^{2+}	−1523

A plot of the values of the enthalpies of hydration of the ions of Table 2.8 against z^2/r_i gives a poor straight line. It may be arranged to be a good straight line by adding the Δ values to r_i. [The reader should try it.]

Table 2.9 Ionic radii, and those given by assuming equation (2.43) to be correct, for some main group cations

Ion	r_i(Born)/pm	r_i/pm	Δ/pm
Li^+	129	76	54
Na^+	165	102	63
K^+	205	138	67
Rb^+	219	152	69
Cs^+	238	167	73
Be^{2+}	110	45	66
Mg^{2+}	142	72	70
Ca^{2+}	170	100	73
Sr^{2+}	187	118	70
Ba^{2+}	208	135	72
Al^{3+}	136	54	79
Ga^{3+}	132	62	70
In^{3+}	150	80	71
Tl^{3+}	152	89	62
Tl^+	199	150	52
Sn^{2+}	176	118	58
Pb^{2+}	183	119	64

Table 2.10 Estimated apparent ionic charges of some main group cations

Ion	Apparent charge/e
Li^+	0.77
Na^+	0.79
K^+	0.82
Rb^+	0.83
Cs^+	0.84
Be^{2+}	1.28
Mg^{2+}	1.42
Ca^{2+}	1.52
Sr^{2+}	1.58
Ba^{2+}	1.61
Al^{3+}	1.92
Ga^{3+}	2.05
In^{3+}	2.19
Tl^{3+}	2.30
Tl^+	0.86
Sn^{2+}	1.64
Pb^{2+}	1.61

Generally, the first method has been used, and Table 2.9 gives the ionic radii, the radii derived from equation (2.43) from the appropriate values of the enthalpies of hydration, and the differences between the two radii. If an amount between 54 and 79 pm (the mean additive amount is 67 ± 8 pm, i.e. $\pm$ one standard deviation) is added to the ionic radii, the Born equation (2.43) gives realistic values for the absolute enthalpies of hydration of the ions of Table 2.8.

The alternative method is to bring the equation (2.43) values into agreement with the estimated values of enthalpies of hydration of the same ions by optimizing their ionic charges. This approach yields the apparent charges given in Table 2.10 for the ions under consideration.

The two methods of bringing the estimated values of the absolute enthalpies of hydration into agreement with the results from equation (2.43) deserve more comment. For example, the lithium cation has an ionic radius of 76 pm and this value gives -918 kJ mol^{-1} when used in equation (2.43). This is a factor of $918/538 = 1.71$ greater negative value than the estimated value. Only by adding 53.5 pm to the ionic radius of the lithium cation does equation (2.43) give a value in agreement with the estimated one. The procedure increases the size of the ion to reduce the effectiveness of its interaction with the solvent. The alternative approach keeps the radii equal to that in the solid state, but optimizes the ionic charge to ensure agreement between the estimated values of the hydration enthalpies and equation (2.43). Both ways indicate that the positive charges on the cations are *delocalized*, and therefore have electrostatic effects somewhat less than those expected for more localized charges.

Worked Problem 2.4

Q Calculate the apparent charge on the Li^+ cation that is necessary to make equation (2.43) produce the value of -538 kJ mol^{-1} for the enthalpy of hydration of the ion.

A With the value of -538 kJ mol^{-1} for the enthalpy of hydration and $r_i = 76$ pm, equation (2.43) becomes:

$$-538 = -\frac{z^2}{76} \times (6.98 \times 10^4)$$

So $z = [(538 \times 76)/69800]^{1/2} = 0.77$.

The very negative value of the estimated enthalpy of hydration of the proton is consistent with a "Born" radius of 63 pm (from equation 2.43). Since that equation overestimates the ionic radius by $\sim$67 pm it follows that a "bare" proton (radius $\sim$0 pm) would be expected to have such a very negative enthalpy of hydration. Studies of the proton and water in the gas phase have shown that the stepwise additions of water molecules in the reaction:

$$H^+(H_2O)_n(g) + H_2O(g) \rightarrow H^+(H_2O)_{n+1}(g) \qquad (2.44)$$

have the standard enthalpy changes given in Table 2.11.

The addition of the first water molecule to the proton causes a much greater release of enthalpy than the following stages and must be associated with the formation of a covalent bond as the ion $H_3O^+(g)$ is produced. The two O–H bonds in the gaseous water molecule have a combined strength of $463 \times 2 = 926$ kJ mol^{-1} and if the amount of 690 kJ mol^{-1} is added the three bonds of the H_3O^+ ion each have strengths of $(926 + 690)/3 = 539$ kJ mol^{-1}. Comparing the bond strengths of O–H (463 kJ mol^{-1}) and F–H (570 kJ mol^{-1}), the value of 539 kJ mol^{-1} for the bonds in H_3O^+ appears to be consistent with the raised electronegativity of the O atom. If the amount 690 kJ mol^{-1} is added to the value of the enthalpy of hydration of the proton, $-1110 + 690 = -420$ kJ mol^{-1}, the residual value of -420 kJ mol^{-1} is equivalent to the enthalpy of hydration of the H_3O^+ ion. When subjected to the Born equation (2.43) this value yields an effective radius of 166 pm. The equation overestimates the radii by 67 pm so the radius of the H_3O^+ ion is 99 pm. This distance is not significantly different from the O–H distance of 96 pm and places the H_3O^+ ion between Li$^+$ and Na$^+$ in terms of ionic radius.

The subsequent additions of water molecules are associated with significantly smaller releases of enthalpy and are consistent with the formation of hydrogen bonds. The total enthalpy change for the production of the ion $H^+(H_2O)_6(g)$ is -1116 kJ mol^{-1}, a value not significantly different from that adopted for the enthalpy of hydration of the proton. Further additions of water molecules to the $H^+(H_2O)_6(g)$ ion are associated with enthalpy changes that are not significantly different from the enthalpy of evaporation of water (44 kJ mol^{-1}) and do not appear to add to the stability of the hydrated proton.

Table 2.12 gives the absolute enthalpies of hydration for some anions. The values are derived from thermochemical cycle calculations using the enthalpies of solution in water of various salts containing the anions and the lattice enthalpies of the solid salts.

The discrepancies between the estimated values of the enthalpies of hydration of cations and the values obtained from equation (2.43) are due to the simple model of the ions adopted by Born. The use of the equations in the reverse manner, *i.e.* accepting the estimated values and adjusting the radius of the ion or its charge to fit, is a proper scientific procedure, one possibly leading to a greater understanding of the nature of ions in aqueous solution.

Table 2.11 Enthalpy changes accompanying the stepwise addition of water molecules to gaseous protons

n	ΔH°/kJ mol^{-1}
0	-690
1	-142
2	-96
3	-71
4	-63
5	-54

Table 2.12 Standard absolute molar enthalpies of hydration of some anions (in kJ mol^{-1})

Ion	$\Delta_{hyd}H^\circ$
F$^-$	-504
Cl$^-$	-359
Br$^-$	-328
I$^-$	-287
ClO$_3^-$	-331
BrO$_3^-$	-358
IO$_3^-$	-446
ClO$_4^-$	-205
OH$^-$	-519
CN$^-$	-341
NO$_3^-$	-316
HCO$_3^-$	-383
CO$_3^{2-}$	-1486
HSO$_4^-$	-362
SO$_4^{2-}$	-1099
PO$_4^{3-}$	-2921

Worked Problem 2.5

Q Estimate the standard enthalpy of hydration of the sulfate ion, given that the lattice enthalpy of sodium sulfate is -1944 kJ mol^{-1} and its standard enthalpy of formation is -1387 kJ mol^{-1}.

A The standard enthalpies of formation of the sodium ion and the sulfate ion are given in Tables 2.3 and 2.2, respectively. The thermochemical cycle shown in Figure 2.11 illustrates the calculation.

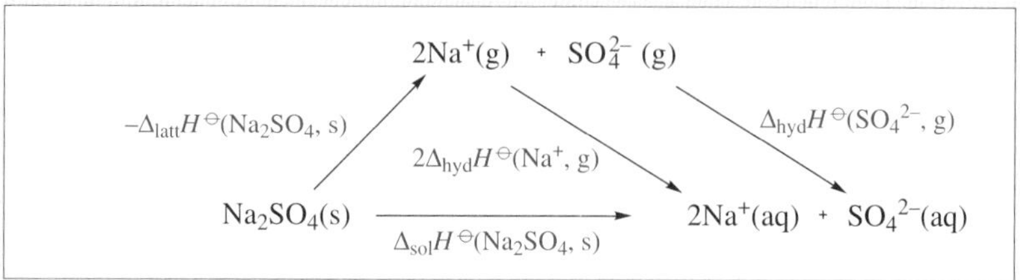

Figure 2.11 A thermochemical cycle for the solution of sodium sulfate

The enthalpy of solution of Na_2SO_4 is calculated from the enthalpies of formation of the products and reactants of the reaction:

$$Na_2SO_4(s) \rightarrow 2Na^+(aq) + SO_4^{2-}(aq) \qquad (2.45)$$

$$\Delta_{sol}H^{\ominus} = \Delta_f H^{\ominus}(SO_4^{2-}, aq) + 2 \times \Delta_f H^{\ominus}(Na^+, aq) - \Delta_f H^{\ominus}(Na_2SO_4, s)$$

$$= -909.3 - (2 \times 240.1) + 1387 = -2.5 \text{ kJ mol}^{-1}.$$

This is also given by the sum:

$$-\Delta_{latt}H^{\ominus}(Na_2SO_4, s) + 2\Delta_{hyd}H^{\ominus}(Na^+, g) + \Delta_{hyd}H^{\ominus}(SO_4^{2-}, g)$$
$$= 1944 - (2 \times 424) + \Delta_{hyd}H^{\ominus}(SO_4^{2-}, g)$$
$$\therefore -2.5 = 1944 - (2 \times 424) + \Delta_{hyd}H^{\ominus}(SO_4^{2-}, g)$$
$$\therefore \Delta_{hyd}H^{\ominus}(SO_4^{2-}, g) = -2.5 - 1944 + (2 \times 424) = -1099 \text{ kJ mol}^{-1}.$$

Lattice enthalpy is defined as the standard molar enthalpy change accompanying the *formation* of one mole of a crystal lattice from its constituent gaseous ions and is quoted as a negative value, since the formation of a lattice is exothermic.

The data of Table 2.12 show that some of the absolute enthalpies of hydration of anions follow the general trend of increasing negative values as the ionic radius decreases and as the ionic charge increases, in line with expectations from equation (2.43). As with cations, equation (2.43) produces values for the enthalpies of hydration of the monatomic halide anions that are slightly different from the thermochemically estimated ones, and which may be brought into agreement by slight alterations in the radii used or in the ionic charges. As is the case with cations, the indication is that anions do not exactly conform to the hard sphere model, but with anions the model is not greatly in error.

As the charge on anions increases there is the expected increase in the negative values of their enthalpies of hydration, *e.g.* compare the values for ClO_4^-, SO_3^{2-} and PO_4^{3-}. The considerably negative value of the enthalpy of hydration of the OH^- ion is consistent with its relatively small thermochemical radius of 152 pm and the capacity that both atoms have of participating in strong hydrogen bonding with water molecules.

The cyanide ion is somewhat larger (187 pm) than the hydroxide ion and does not share its capacity to form strong hydrogen bonds. Its enthalpy of hydration is consequently considerably less negative than that of OH^-. The enthalpies of hydration of NO_3^- (200 pm) and ClO_3^- (208 pm) are very similar, for the two oxoanions possessing a central atom in the $+5$ state. The series of halate(V) ions, ClO_3^-, BrO_3^- (214 pm) and IO_3^- (218 pm), show the opposite trend from that expected from their radii. The explanation lies in the differences in electronegativities of the central halogen atoms. The very electronegative Cl^V drains negative charge away from the ligand oxygen atoms and reduces their capacity to enter into strong hydrogen bonding, whereas in the iodate(V) ion the iodine(V) central atom is not so electronegative and allows the ligand oxygen atoms to participate more fully in hydrogen bonding. The bromate(V) ion has intermediate properties.

If the lattice enthalpy can be estimated from a thermochemical cycle, the value may then be entered into the theoretical equation (*e.g.* Born–Landé) to obtain a value for the sum of the radii of the cation and anion of the compound. This allows an estimate of the anion radius, that of the cation being known. The anion radius obtained in such a manner is called its thermochemical radius.

2.7 Standard Molar Entropies of Hydration of Ions

The calculation of the values for the standard molar entropies of hydration of ions requires some groundwork using the data presented in the following sub-section.

2.7.1 Standard Molar Entropies, $S^{\ominus}$, for Some Aqueous Ions

In this section the standard molar entropies of a small selection of cations and anions are tabulated and the manner of their derivation discussed. The values themselves are required in the calculation of entropies of hydration of ions, discussed in Section 2.7.2.

Cations

The conventional standard entropies of some main group cations are given in Table 2.13, which also includes the standard entropies of the elements.

The conventional standard entropy of an aqueous cation may be calculated from the change in standard entropy for its formation and the standard entropies of the element and that of dihydrogen, as shown in the following example for the aqueous sodium ion.

Table 2.13 Standard molar entropies of some main group elements and the conventional standard molar entropies of their aqueous cations at 25 °C (in $J\,K^{-1}\,mol^{-1}$)

Element/ion	$S^{\ominus}$
Li	29.1
Li$^+$	13.4
Na	51.3
Na$^+$	59.1
K	64.7
K$^+$	103.0
Rb	76.8
Rb$^+$	121.5
Cs	85.2
Cs$^+$	132.9
Be	9.5
Be^{2+}	−129.8
Mg	32.7
Mg^{2+}	−138.6
Ca	41.6
Ca^{2+}	−52.9
Sr	55.0
Sr^{2+}	−29.8
Ba	62.5
Ba^{2+}	9.6
Ra	71.0
Ra^{2+}	54.0
Al	28.3
Al^{3+}	−322.1

Worked Problem 2.6

Q Calculate the conventional standard entropy of the aqueous sodium ion, given that the standard entropy of dihydrogen is $+130.7$ J K^{-1} mol^{-1}.

A The formation reaction for the aqueous sodium ion may be written as:

$$Na(s) + H^+(aq) \rightarrow \tfrac{1}{2} H_2(g) + Na^+(aq) \qquad (2.46)$$

The standard entropy change for the formation reaction is calculated from the values of $\Delta_f G^\circ(Na^+, aq)$ and $\Delta_f H^\circ(Na^+, aq)$ given in Table 2.3:

$$\Delta_f S^\circ = (\Delta_f H^\circ - \Delta_f G^\circ) \times 1000/298.15$$
$$= (-240.1 + 261.9) \times 1000/298.15 = 73.1 \text{ J K}^{-1} \text{mol}^{-1}$$

The standard entropy of formation of an ion is related to the standard entropies of the species participating in the formation reaction:

$$\Delta_f S^\circ(Na^+, aq) = S^\circ(Na^+, aq) + \tfrac{1}{2} S^\circ(H_2, g) - S^\circ(Na, s)$$
$$- S^\circ(H^+, aq) \qquad (2.47)$$

which may be rearranged to give:

$$S^\circ(Na^+, aq) = \Delta_f S^\circ(Na^+, aq) - \tfrac{1}{2} S^\circ(H_2, g) + S^\circ(Na, s)$$
$$+ S^\circ(H^+, aq) \qquad (2.48)$$

$$\therefore S^\circ(Na^+, aq) = 73.1 - 130.7/2 + 51.3 = +59.1 \text{ J K}^{-1} \text{mol}^{-1}.$$

The calculation of the standard entropy of the cation depends upon the *conventional* assumption that the standard entropy of the hydrated proton is zero. A general equation for the standard entropy of an ion $M^{z+}(aq)$, including the term for the standard entropy of the hydrated proton(s), may be written as:

$$S^\circ(M^{z+}, aq) = \Delta_f S^\circ(M^{z+}, aq) - \tfrac{z}{2} S^\circ(H_2, g) + S^\circ(M, s)$$
$$+ z S^\circ(H^+, aq) \qquad (2.49)$$

The last term on the right-hand side is conventionally taken to be zero, so the conventional standard entropies of M^{z+} ions are given by the equation:

$$S^\circ(M^{z+}, aq)^{conv} = \Delta_f S^\circ(M^{z+}, aq) - \tfrac{z}{2} S^\circ(H_2, g) + S^\circ(M, s) \qquad (2.50)$$

Equation (2.49) can then be written as:

$$S^\circ(M^{z+}, aq) = S^\circ(M^{z+}, aq)^{conv} + z S^\circ(H^+, aq) \qquad (2.51)$$

Anions

The conventional standard molar entropies of some anions and some elements are given in Table 2.14.

The conventional standard entropy of an monatomic aqueous anion is calculated from the change in standard entropy for its formation and the standard entropies of the element and that of dihydrogen, as shown in the following worked example for the aqueous chloride ion.

Table 2.14 Conventional standard molar entropies of some elements and the Group 17 anions at 25 °C (in $J\ K^{-1}\ mol^{-1}$)

Element	S°
$H_2(g)$	130.7
$O_2(g)$	205.2
$F_2(g)$	202.8
$Cl_2(g)$	223.1
$Br_2(l)$	152.2
$I_2(s)$	116.1
F^-	−13.8
Cl^-	56.2
Br^-	82.6
I^-	106.5

Worked Problem 2.7

Q Calculate the conventional standard entropy of the aqueous chloride ion, given that of dihydrogen is $+130.7\ J\ K^{-1}\ mol^{-1}$.

A The formation reaction for the aqueous chloride ion may be written as:

$$\tfrac{1}{2}Cl_2(g) + \tfrac{1}{2}H_2(g) \rightarrow H^+(aq) + Cl^-(aq) \qquad (2.52)$$

The standard entropy change for the formation reaction is calculated from $\Delta_f G^\circ(Cl^-, aq)$ and $\Delta_f H^\circ(Cl^-, aq)$, given in Table 2.2:

$$\Delta_f S^\circ = (\Delta_f H^\circ - \Delta_f G^\circ) \times 1000/298.15$$
$$= (-167.2 + 131.2) \times 1000/298.15 = -120.7\ J\ K^{-1}\ mol^{-1}.$$

The standard entropy of formation is related to the standard entropies of the species participating in the formation reaction:

$$\Delta_f S^\circ(Cl^-, aq) = S^\circ(Cl^-, aq) + S^\circ(H^+, aq) - \tfrac{1}{2} S^\circ(H_2, g)$$
$$- \tfrac{1}{2} S^\circ(Cl_2, g) \qquad (2.53)$$

This equation may be rearranged to give:

$$S^\circ(Cl^-, aq) = \Delta_f S^\circ(Cl^-, aq) - S^\circ(H^+, aq) + \tfrac{1}{2} S^\circ(H_2, g)$$
$$+ \tfrac{1}{2} S^\circ(Cl_2, g) \qquad (2.54)$$

$$\therefore S^\circ(Cl^-, aq) = -120.7 + 130.7/2 + 223.1/2 = 56.2\ J\ K^{-1}\ mol^{-1}$$

The value of $S^\circ(H^+, aq)$ is by convention taken to be zero.

A general equation for the calculation of the standard entropy of a monatomic anion is:

$$S^\circ(X^-, aq) = \Delta_f S^\circ(X^-, aq) + \tfrac{1}{2}S^\circ(X_2, g, l\ or\ s)$$
$$+ \tfrac{1}{2}S^\circ(H_2, g) - S^\circ(H^+, aq) \qquad (2.55)$$

The last term on the right-hand side is conventionally taken to be zero, so the conventional standard entropies of X^- ions are given by the equation:

Table 2.15 Absolute standard molar entropies for some aqueous ions (in J K^{-1} mol^{-1})

Ion	S°
Li$^+$	−7.5
Na$^+$	+38.2
K$^+$	+82.1
Rb$^+$	+100.6
Cs$^+$	+112.0
Be^{2+}	−171.6
Mg^{2+}	−180.4
Ca^{2+}	−94.7
Sr^{2+}	−71.6
Ba^{2+}	−32.2
Ra^{2+}	+12.2
Al^{3+}	−384.8
F$^-$	+7.1
Cl$^-$	+77.1
Br$^-$	+103.5
I$^-$	+127.4

The Sackur–Tetrode equation for the molar entropy of a monatomic gas is:

$$S_m = \frac{5R}{2} + R\ln\left[\frac{(2\pi mkT)^{3/2}RT}{N_A h^3 p}\right] \quad (2.61)$$

where m = atomic mass in kg, k = Boltzmann's constant, N_A = Avogadro's constant, h = Planck's constant and p = pressure in pascals (standard pressure = 10^5 Pa).

$$S^{\circ}(X^-, aq)^{conv} = \Delta_f S^{\circ}(X^-, aq) + \tfrac{1}{2}S^{\circ}(X_2, g, l \text{ or } s) + \tfrac{1}{2}S^{\circ}(H_2, g) \quad (2.56)$$

Equation (2.55) can then be written as:

$$S^{\circ}(X^-, aq) = S^{\circ}(X^-, aq)^{conv} - S^{\circ}(H^+, aq) \quad (2.57)$$

Work done with electrochemical cells, with particular reference to the temperature dependence of their potentials, has demonstrated that an accurate value for $S^{\circ}(H^+, aq)$ is −20.9 J K^{-1} mol^{-1}. Table 2.15 gives the absolute molar entropies for the ions under consideration. The values of the absolute standard molar entropies of the ions in Table 2.15 are derived by using the data from Tables 2.13 and 2.14 in equations (2.51) and (2.57).

2.7.2 Absolute Standard Molar Entropies of Hydration of Ions

For the hydration processes $M^{z+}(g) \rightarrow M^{z+}(aq)$ and $X^-(g) \rightarrow X^-(aq)$, the entropy changes depend upon the values of the entropies of the gaseous ions as well as the absolute entropies of the aqueous ions:

$$\Delta_{hyd}S^{\circ}(M^{z+}, g) = S^{\circ}(M^{z+}, aq) - S^{\circ}(M^{z+}, g) \quad (2.58)$$

$$\Delta_{hyd}S^{\circ}(X^-, g) = S^{\circ}(X^-, aq) - S^{\circ}(X^-, g) \quad (2.59)$$

The standard (298.15 K and 10^2 kPa pressure) entropies of monatomic ions in the gas phase may be estimated by using the Sackur–Tetrode equation:

$$S^{\circ} = \tfrac{3}{2}R\ln A_r + 108.9 \quad (2.60)$$

where A_r is the relative atomic mass of the monatomic ion and the second term is the value of a collection of physical constants, the temperature (taken to be 298.15 K) and pressure (taken to be 10^2 kPa = 1 bar.)

The Sackur–Tetrode entropies for one mole of the gaseous ions at 25 °C and 10^2 kPa pressure, *i.e.* with a volume of 24.79 dm^3, $S^{\circ}(M^+, g)$, are given in Table 2.16 for the ions under consideration. The values of the absolute standard entropies of hydration of the ions are given in Table 2.17. The entries are calculated from the data in Tables 2.15 and 2.16 by using equations (2.58) and (2.59).

The values of the absolute standard entropies of hydration are consistent with expectation in that they become less negative as the ion size increases. The lithium cation, being the smallest of Group 1, has the greatest effect on the freedom of movement of the water molecules in its hydration sphere, and causes the greatest reduction in entropy when it becomes hydrated. The fluoride anion, being the smallest of Group 17, has the greatest effect on the freedom of movement of the water molecules in its hydration sphere and causes the greatest reduction in entropy when it becomes hydrated.

Worked Problem 2.8

Q Calculate the value of the absolute standard entropy of hydration of the Mg^{2+} cation.

A The standard entropy of the Mg^{2+} ion in the gas phase is given by Sackur–Tetrode as:

$$S^{\circ}(Mg^{2+}, g) = \tfrac{3}{2}R \ln 24.3 + 108.9 = +148.7 \text{ J K}^{-1} \text{ mol}^{-1}$$

The standard entropy of the hydrated ion is -138.6 J K^{-1} mol^{-1} and the absolute entropy of hydration is given as:

$$\Delta_{hyd} S^{\circ}(Mg^{2+}, g) = -138.6 - 148.7 - (2 \times 20.9) = -329.1 \text{ J K}^{-1} \text{ mol}^{-1}$$

Table 2.16 Absolute standard molar entropies for some gaseous ions (in J K^{-1} mol^{-1})

Ion	S°
Li$^+$	133.1
Na$^+$	148.0
K$^+$	154.6
Rb$^+$	164.4
Cs$^+$	169.9
Be^{2+}	136.3
Mg^{2+}	148.7
Ca^{2+}	154.9
Sr^{2+}	164.7
Ba^{2+}	170.3
Ra^{2+}	176.5
Al^{3+}	150.0
F$^-$	145.6
Cl$^-$	153.4
Br$^-$	163.5
I$^-$	169.3

Inspection of the values for the entropies of hydration of the Group 2 cations in Table 2.17 shows that, with the exception of that for Be^{2+}, the values become less negative as the ionic radius increases. This effect is similar to that observed for the Group 1 cations. The exception of Be^{2+} to the general trend is possibly because of its tendency to have a tetrahedral coordination that causes it to affect fewer molecules of water in the hydration process.

A comparison of the entropies of hydration of Na^+, Mg^{2+} and Al^{3+} shows that they become much more negative as the positive charge on the cation increases. The increasing charge would be expected to be more and more effective in restricting the movement of water molecules, and there would be more water molecules participating in the restricted volume of the hydration spheres of the ions. The decreasing ionic radius in the series amplifies the trend.

2.7.3 The Absolute Standard Molar Entropy of Hydration of the Proton

Table 2.17 Absolute standard molar entropies of hydration for some ions (in J K^{-1} mol^{-1})

Ion	$\Delta_{hyd}S^{\circ}$
Li$^+$	-140.6
Na$^+$	-109.8
K$^+$	-72.5
Rb$^+$	-63.8
Cs$^+$	-57.9
Be^{2+}	-307.9
Mg^{2+}	-329.1
Ca^{2+}	-249.6
Sr^{2+}	-236.3
Ba^{2+}	-202.5
Ra^{2+}	-164.3
Al^{3+}	-534.8
F$^-$	-138.5
Cl$^-$	-76.3
Br$^-$	-60.0
I$^-$	-41.9

Worked Problem 2.9

Q Calculate the value of the absolute entropy of hydration of the proton.

A The standard entropy of a proton gas is given by the Sackur–Tetrode equation:

$$S^{\circ}(H^+, g) = \tfrac{3}{2} R \ln 1 + 108.9 = +108.9 \text{ J K}^{-1} \text{mol}^{-1}$$

and the absolute standard entropy of the hydrated proton is $-20.9 \text{ J K}^{-1} \text{ mol}^{-1}$. The absolute standard entropy of hydration of the proton is therefore $-20.9 - 108.9 = -129.8 \text{ J K}^{-1} \text{ mol}^{-1}$.

The entropy of hydration of the proton has a value intermediate between those of Li^+ and Na^+. This is consistent with the hydrated proton consisting of the hydrated hydroxonium ion, $H_3O^+(aq)$, rather than being the hydrated fundamental particle. The hydrated proton must have a radius intermediate between those of Li^+ and Na^+, consistent with its entropy of hydration.

Summary of Key Points

1. The structure of liquid water was described and compared to that of ice. The importance of hydrogen bonding was indicated.

2. The nature of the hydration of individual ions was described, based upon observations and models.

3. The nature of the hydrated proton was considered in detail.

4. The Gibbs energies of formation, the enthalpies of formation and the standard entropies of ions were described.

5. The enthalpies of hydration of ions were defined and derived. Conventional values were derived and converted into absolute values by using the absolute value of the enthalpy of hydration of the proton.

6. Conventional and absolute values for the entropies of ions were defined and derived.

7. Absolute values for the enthalpies and entropies of hydration of ions were discussed in terms of their sizes and charges.

References

1. F. H. Halliwell and S. C. Nyberg, *Trans. Faraday Soc.*, 1963, **59**, 1126.
2. J. O'M. Bockris and A. K. N. Reddy, *Modern Electrochemistry 1, Ionics*, 2nd edn., Plenum Press, New York, 1998, p. 110.

3. K. R. Asmis, N. L. Pivonka, G. Santambrogio, M. Brümmer, C. Kaposta, D. M. Neumark and L. Wöste, *Science*, 2003, **299**, 1375.
4. E. F. Valeev, E and H. F. Schaefer III, *J. Chem. Phys.*, 1998, **108**, 7197.
5. W. H. Robertson, E. G. Diken, E. A. Price, J.-W. Shin and M. A. Johnson, *Science*, 2003, **299**, 1367.

Further Reading

J. M. Seddon and J. D. Gale, *Thermodynamics and Statistical Mechanics*, RSC Tutorial Text No 10, Royal Society of Chemistry, Cambridge, 2001.
Y. Marcus, *Ion Solvation*, Wiley, New York, 1985.
H. L. Friedman and C. V. Krishnan, *Thermodynamics of Ion Hydration* in F. Franks (ed.), *Water, A Comprehensive Treatise*, vol. 3, *Aqueous Solutions of Simple Electrolytes*, chap. 1, Plenum Press, New York, 1973.
L. G. Hepler and E. M. Woolley, *Hydration Effects and Acid-Base Equilibria* in F. Franks (ed.), *Water, A Comprehensive Treatise*, vol. 3, *Aqueous Solutions of Simple Electrolytes*, chap. 3, Plenum Press, New York, 1973.
J. O'M. Bockris and A. K. N. Reddy, *Modern Electrochemistry 1, Ionics*, 2nd edn., Plenum Press, New York, 1998.
Y. Marcus, *Ionic Radii in Aqueous Solutions*, in *Chem. Rev.*, 1988, **88**, 1475.

Problems

2.1. Estimate the standard enthalpy of hydration of the cyanide ion, given that the lattice enthalpy of potassium cyanide is -692 kJ mol^{-1} and its standard enthalpy of formation is -113 kJ mol^{-1}.

2.2. Estimate the standard enthalpy of hydration of the hydroxide ion, given that the lattice enthalpy of potassium hydroxide is -802 kJ mol^{-1} and its standard enthalpy of formation is -424.6 kJ mol^{-1}.

2.3. Estimate the standard enthalpy of hydration of the hydrogen sulfate ion, given that the lattice enthalpy of sodium hydrogen sulfate is -784 kJ mol^{-1} and its standard enthalpy of formation is -1125.5 kJ mol^{-1}.

2.4. Estimate the standard enthalpy of hydration of the chlorate(VII) ion, given that the lattice enthalpy of sodium chlorate(VII) is -643 kJ mol^{-1} and its standard enthalpy of formation is $-383.3 \text{ kJ mol}^{-1}$.

2.5. Calculate the value of the absolute standard entropy of hydration of the Be^{2+} cation.

2.6. Calculate the value of the absolute standard entropy of hydration of the Sr^{2+} cation.

3

Acids and Bases; Forms of Ions in Aqueous Solution; Ion Hydrolysis & Compound Solubility

In this chapter the **Brønsted–Lowry theory of acids and bases** in aqueous solutions is described. **Ion hydrolysis**, the reaction between an ion and solvent water molecules that governs the forms of ions in solution, is described and explained. The factors that govern the **solubility of inorganic compounds** in water are discussed.

Aims

By the end of this chapter you should understand:

- The Brønsted–Lowry definitions of acids and bases
- The factors governing acidity
- The hydrolysis of ions in aqueous solutions
- The factors which influence the forms of ions in aqueous solution
- The criteria that determine whether or not compounds dissolve in water
- The roles of hydration enthalpies and entropies in determining the solubilities of ionic compounds

3.1 Acids and Bases in Aqueous Solutions

The subject of acids and bases is very extensive. The discussion in this book is restricted to the definitions of acids and bases in aqueous solutions and their applications to the nature of ions in aqueous solutions and their stabilities. The two main definitions are those accredited to

Table 3.1 Definitions of acids and bases

	Brønsted/ Lowry	Lewis
Acid	Proton donor	Electron-pair acceptor
Base	Proton acceptor	Electron-pair donor

Brønsted–Lowry acids and bases are also Lewis acids and bases if the proton transferred in the acid–base reaction is regarded as an electron-pair acceptor and the base is regarded as an electron-pair donor. For example, in the equilibrium:

$$H_3O^+(aq) + OH^-(aq)$$
$$\rightleftharpoons 2H_2O(l) \qquad (3.2)$$

the hydroxonium cation furnishes a proton which is accepted by the hydroxide anion. In electron-pair terms, the proton "accepts" an electron pair "donated" by the hydroxide ion in the formation of the H–O bond in the newly formed water molecule.

The molar concentration of a substance in solution is represented by the symbol c, with units of mol dm^{-3}.

(i) Brønsted and Lowry and (ii) G. N. Lewis. They are the ones that are currently in most use by chemists, but the first one is specifically relevant to aqueous solutions. The definitions are summarized in Table 3.1.

3.2 Brønsted and Lowry Acids and Bases

A **Brønsted–Lowry acid** is defined as a **proton donor,** a Brønsted–Lowry **base** as a **proton acceptor.** The definitions apply generally to **protic** systems: those in which **proton transfers** can occur. A general equation expressing proton transfer in aqueous solution is:

$$AH(aq) + B(aq) \rightleftharpoons A^-(aq) + BH^+(aq) \qquad (3.1)$$

where AH represents a general acid with a dissociable proton and B represents a general base which can accept a proton. In the reverse process, where the proton is donated by the BH$^+$ ion to the anion A$^-$, BH$^+$ is called the **conjugate acid** of the base B and the anion A$^-$ is called the **conjugate base** of the acid AH.

The strength of a Brønsted–Lowry acid is quantified in terms of the magnitude of the equilibrium constant for the ionization reaction in which the solvent acts as the base, *e.g.* for the aqueous system, the general reaction is:

$$AH(aq) + H_2O(l) \rightleftharpoons H_3O^+(aq) + A^-(aq) \qquad (3.3)$$

The equilibrium constant (the acid dissociation constant), K_a, is given by:

$$K_a = \frac{a_{H_3O^+} a_{A^-}}{a_{HA} a_{H_2O}} \qquad (3.4)$$

where the a terms represent the activities of the sub-scripted species. The activity of a substance is given by:

$$a = \gamma c / a^\ominus \qquad (3.5)$$

the product of the activity coefficient, γ, and the molar concentration, c, of the substance divided by the standard activity, $a^\ominus$, equal to 1 mol dm^{-3}.

In the case of the solvent (water, concentration = 55.5 mol dm^{-3}), its standard activity is usually taken to be 1. For solvents, the definition of activity is the ratio of its vapour pressure when acting as a solvent, p, divided by the vapour pressure of the pure solvent, $p^\ominus$, taken as the standard state:

$$a_{solvent} = p/p^\ominus \qquad (3.6)$$

For dilute solutions $p \approx p^\ominus$ and $a_{solvent} = 1$.

Thus the equilibrium constant for acid dissociation may be written as:

$$K_a = \frac{a_{H_3O^+} a_{A^-}}{a_{HA}} \tag{3.7}$$

Activity coefficients are introduced into equations such as (3.7) to account for the non-ideal behavior of solutes and solvents. The equation then becomes:

$$K_a = \frac{\gamma_+ c_{H_3O^+} \gamma_- c_{A^-}}{\gamma_{HA} c_{HA}} \tag{3.8}$$

Because the individual ionic activity coefficients cannot be measured, they are replaced in equation (3.8) by the **mean ionic activity coefficient,** the geometric mean of the two individual coefficients:

$$\gamma_\pm = (\gamma_+ \gamma_-)^{1/2} \tag{3.9}$$

$$K_a = \frac{\gamma_\pm c_{H_3O^+} c_{A^-}}{\gamma_{HA} c_{HA}} \tag{3.10}$$

In dilute solutions the activity coefficients of the dissolved substances may be taken to be 1.0. The equilibrium constant may, under those conditions, be represented by the equation:

$$K_a = \frac{[H_3O^+][A^-]}{[HA]} \tag{3.11}$$

where the quantities in square brackets are the concentrations of the indicated species compared to the standard concentration of 1 mol dm^{-3}.

The convention of expressing the concentration of a substance as a ratio (actual concentration/standard concentration) has the effect of making expressions for equilibrium constants dimensionless. They do not have units.

Table 3.2 Some representative pK_a values for acids at 25 °C; the negative values are only very approximate

Acid	pK_1	pK_2	pK_3
HF	3.2		
HCl	−6.3		
HBr	−8.7		
HI	−9.3		
HCN	9.21		
H_2SO_4	<0	1.99	
HNO_3	−1.37		
H_3PO_4	2.16	7.21	12.32
H_2CO_3	3.7 or 6.38[a]	10.32	
H_3BO_3	9.27		
$HClO_4$	<< 0		

[a] The difference is discussed in Section 3.2.3.

There is a very large range of values of K_a, so it is more convenient to express acid strengths as pK_a values ($pK_a = -\log_{10} K_a$). Some values, exemplifying the extensive range, are given in Table 3.2.

The negative pK_1 values of the "strong" acids, HCl, HClO$_4$, H$_2$SO$_4$ and HNO$_3$, indicate that in aqueous solution they are, for all practical purposes, completely dissociated. Sulfuric acid is **dibasic** or **diprotic**, *i.e.* it has two dissociable protons, and although the first dissociation is complete in aqueous solution, the second dissociation ($pK_2 = 1.99$) indicates that the HSO$_4^-$ ion is not a particularly strong acid. Phosphoric acid is **tribasic (triprotic)**, and exhibits a general pattern for **polybasic (polyprotic)** acids of pK_a values that increase by about five units for successive dissociations.

The data given in Table 3.2 may be interpreted for a general acid H–A, using thermochemical cycles, in terms of the enthalpy changes accompanying the reactions:

$$HA(aq) \rightarrow HA(g) \tag{3.12}$$

$$HA(g) \rightarrow H(g) + A(g) \tag{3.13}$$

$$H(g) + A(g) \rightarrow H^+(g) + A^-(g) \tag{3.14}$$

$$H^+(g) + A^-(g) \rightarrow H^+(aq) + A^-(aq) \tag{3.15}$$

where A is the residue of the HA molecule after dissociation of the hydrogen atom.

Three observations are discussed in this text: (i) the weak acidity of HF compared with the other halogen halides; (ii) the weak acidity of HCN; and (iii) the increasing acid strength with the oxidation state of the central element in the oxoacid series H$_3$BO$_3$, H$_2$CO$_3$ and HNO$_3$ of the second period of the Periodic Table, and H$_3$PO$_4$, H$_2$SO$_4$ and HClO$_4$ of the third period.

3.2.1 The Weak Acid Strength of HF

Hydrogen fluoride in aqueous solution is a weak acid, characterized by its pK_a value of 3.2. By comparison, the other hydrogen halides are extremely strong acids in aqueous solution; all three are fully dissociated in dilute solution, and their pK_a values may be estimated by thermochemical cycle calculations. The thermochemical cycle shown in Figure 3.1 represents the various processes as the aqueous hydrogen halide, HX, is converted to a solution containing hydrated protons and hydrated halide ions. The enthalpy of acid dissociation of the HX(aq) compound is given by:

$D(HX, g)$ is the dissociation enthalpy of HX(g).
$I(H)$ is the ionization energy of H(g).
$E(X)$ is the electron attachment energy of X(g).
The 6.2 kJ mol^{-1} required to convert $I(H)$ into an enthalpy change is omitted. It would be cancelled out by the -6.2 kJ mol^{-1} that is needed to convert $E(X)$ to an enthalpy change.

$$\Delta_{diss}H^\circ(HX, aq) = -\Delta_{hyd}H^\circ(HX, g) + D(HX, g) + I(H) + E(X)$$
$$+ \Delta_{hyd}H^\circ(H^+, g) + \Delta_{hyd}H^\circ(X^-, g) \tag{3.16}$$

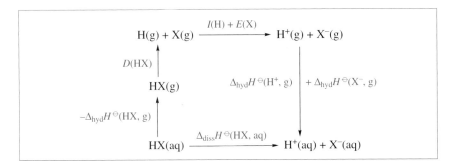

Figure 3.1 A thermochemical cycle for the acid dissociation of HX molecules

Values for the quantities in equation (3.15) for $X = F$ and Cl are given in Table 3.3.

Table 3.3 Calculations of enthalpies of dissociation for HF and HCl

	F	Cl
$-\Delta_{hyd}H^{\circ}$ (HX, g)	48	18
D(HX, g)	570	432
I(H)	1312	1312
E(X)	−328	−349
$\Delta_{hyd}H^{\circ}$ (H^{+}, g)	−1110	−1110
$\Delta_{hyd}H^{\circ}$ (X^{-}, g)	−504	−359
$\Delta_{diss}H^{\circ}$ (HX, aq)	−12	−56

The values of $\Delta_{diss}H^{\circ}$(HX, aq) for both acids are considerably negative and would contribute those negative amounts to the Gibbs energy changes for the dissociation processes. Only if the changes in Gibbs energy for the dissociations are positive would the acids be weak. The conclusion may only be reached by considering the entropy changes that occur when the compounds dissociate. The $T\Delta_{diss}S^{\circ}$ terms for HF and HCl are −29 and −13 kJ mol^{-1}, respectively, and these produce values of $\Delta_{diss}G^{\circ}$(HF, aq) $= +17$ kJ mol^{-1} and of $\Delta_{diss}G^{\circ}$(HCl, aq) $= -43$ kJ mol^{-1}. Using the relationship $-\Delta G^{\circ} = RT\ln K_a$, the p$K_a$ values for the two acids are 3.0 and −7.5, respectively. There are three enthalpy factors that decide the difference between the enthalpies of dissociation of HF and HCl. HF in aqueous solution is stabilized by extensive hydrogen bonding and the H–F bond strength is high. These two factors are offset somewhat by the very negative value of the hydration enthalpy of the fluoride ion. HCl is not significantly hydrogen bonded in solution, and it is more easily dissociated in the gas phase. These two factors are offset by a less negative

$T\Delta_{hyd}S^{\circ}(F^{-}, g) - T\Delta_{hyd}S^{\circ}(Cl^{-}, g)$
$= T \times (-62.2) = -18.5 \text{ kJ mol}^{-1}$
$T\Delta_{diss}S^{\circ}(HF, aq) - T\Delta_{diss}S^{\circ}(HCl, aq) = -29-(-13) = -16 \text{ kJ mol}^{-1}$

value of the hydration enthalpy of the chloride ion. The deciding factor in making the Gibbs energy of dissociation for HF positive is the overall entropy change. The main differences in entropy changes between the two acids lies with the entropies of hydration of the two negative ions; $\Delta_{hyd}S^{\circ}(F^{-}, g) = -138.5 \text{ J K}^{-1} \text{ mol}^{-1}$ and $\Delta_{hyd}S^{\circ}(Cl^{-}, g) = -76.3 \text{ J K}^{-1}$ mol^{-1}, the difference of $-62.2 \text{ J K}^{-1} \text{ mol}^{-1}$ amounting to a difference of $-18.5 \text{ kJ mol}^{-1}$ in the $T\Delta_{diss}S^{\circ}$ terms.

A further complication with HF in aqueous solution is that infrared spectroscopic studies[1] have shown the presence of a high concentration of H_3O^+ ions, although they are highly associated with fluoride ions by hydrogen bonding: $H_3O^+ \cdots F^-$. It is argued that this association has a drastic effect on the activity coefficient of the hydrogen ion, and that this results in the high value of pK_a for HF.

3.2.2 The Weak Acid Strength of HCN

The cyanide ion is sometimes described as a pseudo-halide ion, *i.e.* it behaves much like a halide ion. It has a radius of 187 pm, not very different from the 181 pm radius of the chloride ion, yet HCl is a very strong acid and HCN is very weak, the difference between the two pK_a values being about 16 units. The thermochemical cycle shown in Figure 3.1 is also applicable to the HCl/HCN comparison, and appropriate data are given in Table 3.4.

Table 3.4 Calculations of enthalpies of dissociation for HCl and HCN

	Cl	*CN*
$-\Delta_{hyd}H^{\circ}$ (HX, g)	18	26
D(HX, g)	432	528
I(H)	1312	1312
E(X)	-349	-373
$\Delta_{hyd}H^{\circ}$ (H$^+$, g)	-1110	-1110
$\Delta_{hyd}H^{\circ}$ (X$^-$, g)	-359	-341
$\Delta_{diss}H^{\circ}$ (HX, aq)	-56	+42

The difference in the two values for enthalpy of dissociation of $56 + 42 = 98 \text{ kJ mol}^{-1}$ is equivalent to a difference in pK_a of 17 units in favour of HCl being the stronger acid. The factors that contribute to reducing the value for HCN are its high H−CN dissociation enthalpy, its relatively large negative electron attachment enthalpy, and its relatively less negative value for the enthalpy of hydration of the ion.

3.2.3 The Dependence of Acid Strengths of Oxoacids on the Oxidation State of the Central Element

Two comparisons are discussed in this section: (a) the acid strengths of the oxoacids of the second period, H_3BO_3, H_2CO_3 and HNO_3, and (b) those of the oxoacids of the third period, H_3PO_4, H_2SO_4 and $HClO_4$.

(a) Boric acid, H_3BO_3, is better formulated as boron trihydroxide, $B(OH)_3$, and is a very weak acid with a pK_1 of 9.27. Even in a solution of 1 mol dm^{-3} NaOH the acid is only dissociated as far as the removal of the first proton. Carbonic acid, H_2CO_3, is another hydroxoacid and is better formulated as $OC(OH)_2$. Although still a weak acid, it is considerably stronger than boric acid. In aqueous solution its chemistry is complicated by the presence of hydrated carbon dioxide, CO_2(aq), in addition to the undissociated acid. The first dissociation of carbonic acid is governed by the equation:

$$K_1 = \frac{[H^+][HCO_3^-]}{[H_2CO_3]} \qquad (3.17)$$

with K_1 having a value of 4.16×10^{-7} ($pK_1 = 6.38$) if the $[H_2CO_3]$ term represents all the neutral aqueous carbon dioxide. If the two forms of the aqueous oxide are distinguished and the real value of the undissociated carbonic acid used in equation (3.20), the value of K_1 becomes 2×10^{-4} ($pK_1 = 3.7$). This latter value is a better representation of the acid strength of carbonic acid, and shows it to be considerably stronger than boric acid. Nitric(V) acid may be formulated as a hydroxoacid, $O_2N(OH)$, and is a very strong acid. In the series $B(OH)_3$, $OC(OH)_2$, $O_2N(OH)$, the oxidation state of the central atom increases, *i.e.* B^{III}, C^{IV}, N^V, and is the main cause of the increased acid strength, the more oxidized central atom having greater electronegativity and attraction for the electrons that are used to bind the oxygen atoms to it. The greater the share of these bonding electrons is, the weaker the O–H bond becomes, and this facilitates the dissociation of a proton. This is a generally observed phenomenon. The pK_1 values of the three acids are 9.3, 3.7 and −1.37, decreasing by 4.6 and 5.07 units for each successive increase in the oxidation number of the central atom.

(b) In the third period oxoacids, H_3PO_4, H_2SO_4 and $HClO_4$, there is also an increase of acid strength as the oxidation state of the central atom increases ($+5 \rightarrow +6 \rightarrow +7$). The relatively strong phosphoric acid, $OP(OH)_3$, has a pK_1 value of 2.16, but in sulfuric acid, $O_2S(OH)_2$, the S^{VI} central atom has a greater effect on the O–H bond strength, to the extent of making the pK_1 value negative. This trend continues with chloric(VII) acid having a very negative pK_1 value. Estimates of negative pK values are not very reliable, and the best evidence for considering

$HClO_4$ to be a stronger acid than sulfuric acid is that in a mixture of the two compounds the sulfuric acid molecule is protonated:

$$HClO_4 + H_2SO_4 \rightarrow H_3SO_4^+ + ClO_4^- \tag{3.18}$$

Pauling stated two rules for the magnitudes of the pK values of hydroxoacids:

1. For a given hydroxoacid the successive pK values increase by 5 units.
2. For a hydroxoacid $O_mE(OH)_n$, the acid strength depends on the values of m, as shown in Table 3.5.

Table 3.5   The acid strengths of hydroxoacids, $O_mE(OH)_n$

m	pK_1	Acid strength	Examples
0	≥ 7	Very weak	$B(OH)_3$, 9.3; $Cl(OH)$, 7.5
1	~ 2	Weak	$OCl(OH)$, ~2; $OS(OH)_2$, 2.0; $OP(OH)_3$, 2.18
2	−3	Strong	$O_2Cl(OH)$, $O_2S(OH)_2$, $O_2N(OH)$
3	−8	Very strong	$O_3Cl(OH)$

For each increase in the value of m by one, the oxidation state of the central atom rises by two units, with the consequent increase in acid strength as the O–H bond strength decreases. The increase in acid strength with the electronegativity of the central atom is dominated by the number of formal element–oxygen double bonds. These concentrate electronic charge in the regions far from the hydroxo group that is participating in the dissociation process. In contrast, the many hydroxide compounds of electropositive elements, e.g. Na and Ca, are basic; they dissociate into cations of the electropositive elements and hydroxide ions. Those soluble in water are strong alkalis.

3.3   pH and the pH Scale

The autoprotolysis of liquid water was introduced in Chapter 1. The process is one of proton transfer between two molecules, as indicated by the equilibrium:

$$2H_2O(l) \rightleftharpoons H_3O^+(aq) + OH^-(aq) \tag{3.19}$$

The positive ion is a hydrated hydrogen ion and the negative ion is a water molecule minus one hydrogen ion. The water molecules act equally as acids and bases; this type of behaviour is termed amphiprotic. The extent of the autoionization is very slight. The autoprotolysis constant,

K_w (the equilibrium constant for the process described by equation 3.19), has a value of 1.0×10^{-14} (298.15 K):

$$K_w = a_{H_3O^+} a_{OH^-} \qquad (3.20)$$

Expressed in logarithmic form this becomes:

$$pK_w = pH + pOH = 14 \qquad (3.21)$$

where $pH = -\log_{10} a_{H_3O^+}$ and $pOH = -\log_{10} a_{OH^-}$.

Worked Problem 3.1

Q If the pH of a solution is 6, calculate the value of pOH.

A $pH + pOH = 14$, so $pOH = 14 - pH = 14 - 6 = 8$

Pure water containing no dissolved gases possesses equal concentrations of hydronium and hydroxide ions, so that pH = pOH = 7.0 at 298.15 K. This is defined as a **neutral solution**. This allows a practical definition of acidic and basic behaviour. Any substance dissolved in water which produces a pH below 7 is termed an acid and any substance dissolved in water which produces a pH greater than 7 is called a base.

Substances typical of acids and bases are, respectively, HCl and NaOH. Hydrogen chloride dissolves in water with practically complete dissociation into hydrated protons and hydrated chloride ions. Sodium hydroxide dissolves in water to give a solution containing hydrated sodium ions and hydrated hydroxide ions. Table 3.6 gives values of the mean ionic activity coefficients, $\gamma_\pm$, at different concentrations and indicates the pH values and those expected if the activity coefficients are assumed to be unity.

Table 3.6 Mean ionic activity coefficients and pH values for some aqueous solutions of HCl and NaOH

	Molar concentration, c	$\gamma_\pm$	Activity, a	$pH = -\log_{10} a$	$-\log_{10} c$
HCl	0.001	0.966	0.000966	3.02	3.0
	0.01	0.905	0.00905	2.04	2.0
	0.1	0.796	0.0796	1.10	1.0
	1.0	0.809	0.809	0.09	0.0
NaOH	0.01	0.899	0.00899	11.95	12.0
	0.1	0.766	0.0766	12.88	13.0
	1.0	0.679	0.679	13.83	14.0

The non-ideality of aqueous solutions of substances producing hydrated ions is an indication of the complexity of such systems. In addition to primary and secondary hydration shells, ions are surrounded by a number of ions of opposite charge and this causes the ions to behave in a non-ideal manner. The clusters of anions around cations and those of cations around anions have a transitory existence. There is also rapid exchange of water between the primary and secondary hydration shells of the ions and with water molecules in the bulk solution.

The deviations from ideality expressed by the right-hand pH column of Table 3.6, are not great, and are ignored unless stricter accuracy is required.

The pH values produced by 1 mol dm^{-3} solutions of HCl and NaOH, *i.e.* 0 and 14 respectively, define the practical range in which activity coefficients of H$^+$(aq) and OH$^-$(aq) may be ignored for general purposes, and subsidiary definitions of pH and pOH may be used: pH $= -\log_{10}[H^+]$ and pOH $= -\log_{10}[OH^-]$. The square brackets indicate the molar concentration of the species enclosed as a ratio to the standard molar concentration of 1 mol dm^{-3}.

3.4 Factors Influencing Acidic and Basic Behaviour in Aqueous Solutions

Na$_2$O is a basic oxide; it dissolves in water to give an alkaline solution. SO$_3$ is an acidic oxide; it dissolves in water to give an acid solution. Al$_2$O$_3$ dissolves in a solution of NaOH to give the AlO$_2^-$(aq) ion; the oxide is reacting in an acidic manner. The compound also is soluble in acids in which the Al^{3+}(aq) ion is formed; the oxide is acting as a base. Compounds showing both acidic and basic properties are called amphoteric. Al$_2$O$_3$ is an amphoteric oxide.

Some binary hydrides (*e.g.* those of Groups 16 and 17) behave as acids in aqueous solution, but the majority of acids are oxoacids derived from acidic oxides. This discussion is restricted to the factors influencing the production of acids or bases when oxides dissolve in water. An oxide can be acidic, **amphoteric** (*i.e.* acidic or basic depending upon conditions) or basic.

A general understanding of oxide behaviour may be achieved by a consideration of the E–O–H grouping in the hydroxide produced by the reaction of an oxide with water (E representing any element):

$$EO + H_2O \rightarrow E(OH)_2 \tag{3.22}$$

The compound behaves as an acid if the O–H bond is broken in a heterolytic manner, *i.e.* the oxygen atom retains both of the bonding electrons:

$$E–O–H \rightarrow E–O^-(aq) + H^+(aq) \tag{3.23}$$

If the E–O bond breaks in a similar heterolytic manner, the compound exhibits basic behaviour:

$$E–O–H \rightarrow E^+(aq) + OH^-(aq) \tag{3.24}$$

Intermediate compounds adopt amphoteric behaviour. If E is a relatively large and electropositive, *e.g.* Na, the E–O–H arrangement is likely to be ionic, E$^+$OH$^-$, so that the hydrated hydroxide ion is produced when the compound dissolves in water. If E is relatively small, electronegative and in a high oxidation state, *e.g.* as the chlorine(VII) atom in O$_3$Cl–O–H, the E–O bond is likely to be the stronger bond, thus favouring heterolysis to give the ions EO$^-$(aq) and H$^+$(aq). In Al(OH)$_3$, which is an intermediate case with a central element in the $+3$ oxidation state, both heterolytic processes can occur:

$$3OH^-(aq) + Al^{3+}(aq) \rightleftharpoons Al(OH)_3(s) \rightleftharpoons H_3O^+(aq) + AlO_2^-(aq) \quad (3.25)$$

Basic dissociation is encouraged by the presence of an acidic solution (*i.e.* pH < 7) and, conversely, acidic dissociation is encouraged by the presence of a basic solution (*i.e.* pH > 7).

Sodium hydroxide dissolves in water to give $OH^-(aq)$ ions, the strongest base which can exist in an aqueous system. Chloric(VII) acid (perchloric acid) dissociates practically completely in dilute solution to give $H_3O^+(aq)$ ions, which represent the strongest acid which can exist in an aqueous system. The amphoteric behaviour of aluminium is noticed in a series of hydrated salts containing the Al^{3+} ion and in compounds such as $NaAlO_2$, which contains the aluminate(III) ion, AlO_2^-.

3.5 Forms of Ions in Aqueous Solution; Hydrolysis

The molecular form of any ion in aqueous solution depends largely upon the size and charge (oxidation state) of the central atom. These two properties influence the extent of any interaction with the solvent water molecules. A large singly charged ion is likely to be simply hydrated. In the case of a transition metal ion, M^{n+}, there are commonly six water molecules of hydration actually bonded to the metal ion by coordinate (dative) bonds. This set of water molecules constitutes the primary hydration shell. Such ions are surrounded by more loosely hydrogen-bonded molecules of water that form the secondary hydration shell. The interaction of gaseous ions with water to give their hydrated versions causes the liberation of enthalpy as the enthalpy of hydration, $\Delta_{hyd}H^{\ominus}$, which is *negative*, and represents a major contribution to their thermodynamic stability.

Forms of ions other than the simply hydrated ones may be understood in terms of the increasing interaction between the central ion and its hydration sphere as its oxidation state increases and its size decreases. This interaction is a chemical change known as **hydrolysis**. The first stage of hydrolysis may be written as:

$$E(H_2O)^{n+}(aq) \rightleftharpoons EOH^{(n-1)+}(aq) + H^+(aq) \quad (3.26)$$

implying that an element–oxygen bond is formed with the release of a proton into the solution. As the value of n (the oxidation state of E) increases, the next stage of hydrolysis would be expected to occur:

$$E(H_2O)OH^{(n-1)+}(aq) \rightleftharpoons E(OH)_2^{(n-2)+}(aq) + H^+(aq) \quad (3.27)$$

There is then the possibility that water could be eliminated from the two OH^- groups:

$$E(OH)_2^{(n-2)+}(aq) \rightleftharpoons EO^{(n-2)+}(aq) + H_2O(l) \quad (3.28)$$

Further hydrolyses involving more pairs of water molecules would yield the oxo ions $EO^{(n-4)+}$, $EO^{(n-6)+}$ and $EO^{(n-8)+}$.

Another possibility is that a water molecule could be eliminated between two hydrolysed ions to give a dimeric product:

$$2EOH^{(n-1)+}(aq) \rightleftharpoons EOE^{2(n-1)+}(aq) + H_2O(l) \qquad (3.29)$$

Further hydrolysis of the dimeric product could yield ions such as $O_3EOEO_3^{2(n-7)+}$. In some cases, further condensations could occur to give polymeric ions.

Examples of these forms of ions are to be found in the chemistry of the transition elements and the main group elements that can exist in higher oxidation states. As may be inferred from equations (3.26) and (3.27), alkaline conditions encourage hydrolysis, so that the form an ion takes depends on the pH of the solution. Highly acid conditions tend to depress the tendency of an ion to undergo hydrolysis. Table 3.7 contains some examples of ions of different form, the form depending upon the oxidation state of the central element, vanadium.

Table 3.7   Some ionic forms of vanadium in solution

Oxidation state of V	Form in acid solution	Form in alkaline solution
II	V^{2+}	VO (insoluble)
III	V^{3+}	V_2O_3 (insoluble)
IV	VO^{2+}	$V_2O_5^{2-}$
V	VO_2^+	VO_4^{3-}

In acidic solution the $+2$ and $+3$ oxidation states are simple hydrated ions, the $+4$ and $+5$ states being oxocations. In alkaline solution the $+2$ and $+3$ states form neutral insoluble oxides (the lattice energies of the oxides giving even more stability than any soluble form in these cases), and the $+4$ and $+5$ states exist as oxyanions (dimeric in the $+4$ case).

Simple cations such as $[Fe(H_2O)_6]^{3+}$ undergo a certain amount of primary hydrolysis, depending upon the pH of the solution. The ion is in the hexaaqua form only at pH values lower than 2.0. Above that value the hydroxopentaaquairon(III) ion, $[Fe(H_2O)_5OH]^{2+}$, is predominant. Further increase in the pH of the solution causes more hydrolysis, until a complex solid material sometimes described erroneously as iron(III) hydroxide is precipitated. The solid does not have the formula $Fe(OH)_3$, but contains iron(III) oxohydroxide (FeOOH) and iron(III) oxide in various states of hydration, $Fe_2O_3.xH_2O$.

With some ions the extent of polymer formation is high, even in acidic solution, e.g. vanadium(V) can exist as a polyvanadate ion, $V_{10}O_{28}^{6-}$.

The above ideas are not limited to species with central metal ions. They apply to the higher oxidation states of non-metallic elements. Many simple anions do exist with primary hydration spheres in which the positive ends of dipoles are attracted to the central negative charge. Table 3.8 gives examples of ions that may be thought about in terms of the hydrolysis of "parent" hypothetical hydrated ions.

The hydrolysis products from the hypothetical Si^{4+} ion are the many and varied polymeric silicate ions which exist as chains, double chains, and sheets, culminating in the insoluble neutral three-dimensional arrays with the formula SiO_2.

Table 3.8 Hydrolysis products of some ions in high oxidation states

Hypothetical bare ion	Hydrolysis product(s)
S^{4+}	SO_3^{2-}
S^{6+}	SO_4^{2-} and $S_2O_7^{2-}$
Cl^+	ClO^-
Cl^{3+}	ClO_2^-
Cl^{5+}	ClO_3^-
Cl^{7+}	ClO_4^-

3.6 Solubilities of Ionic Compounds in Water

The various ways in which solubility data are presented are discussed in Section 1.1. This section consists of a description of the factors governing the solubility of inorganic compounds in water. The discussion is restricted mainly to the Group 1 halides with some concluding generalizations.

Worked Problem 3.2

Q Calculate the molar concentrations in mol dm^{-3} of the sodium halides (that for NaCl is calculated in Worked Problem 1.1), given that their saturated solutions at 25 °C have the compositions:

The molar concentration of NaCl in its saturated solution at 25 °C is 5.42 mol dm^{-3}.

	NaF	NaBr	NaI
Solubility/g/100 g water	4.13	94.6	184
Density of solution/kg m^{-3}	1038	1542	1918

A The numbers of moles of the compounds in 100 g of water are obtained by dividing the mass of the compounds by their relative formula masses. The volumes of the solutions in dm^3 are obtained by dividing the total mass by the density. Dividing the numbers of moles of the halides by the volumes of their solutions gives the required molar concentrations. The answers are:

	NaF	NaBr	NaI
Moles	0.098	0.918	1.23
Volume of solution/cm^3	100.3	126.2	148.1
Molar concentration/mol dm^{-3}	0.098	7.27	8.31

The results given in the text and the worked problem indicate that the solubilities of the sodium halides vary from the low value for the fluoride to the much higher values of the other halides.

This variation is associated with the Gibbs energies of solution, $\Delta_{sol}G^{\circ}$, of the compounds. These, and the values of the enthalpies of solution, $\Delta_{sol}H^{\circ}$, and the entropy changes of solution, $\Delta_{sol}S^{\circ}$, when the compounds dissolve in water, are given in Table 3.9.

Table 3.9   Standard Gibbs energy, enthalpy and entropy changes of solution for the sodium halides

Compound	$\Delta_{sol}G^{\circ}$ /kJ mol^{-1}	$\Delta_{sol}H^{\circ}$ /kJ mol^{-1}	$\Delta_{sol}S^{\circ}$ /J K^{-1} mol^{-1}
NaF	+5.6	+3.8	−6.0
NaCl	−9.0	+3.9	+43.3
NaBr	−16.8	−0.6	+54.3
NaI	−27.4	−7.5	+71.8

Worked Problem 3.3

Q   Use the data in Table 3.9 to calculate the value of $\Delta_{sol}S^{\circ}$ from the values of $\Delta_{sol}H^{\circ}$ and $\Delta_{sol}G^{\circ}$ for sodium bromide at 298.15 K.

A $\Delta_{sol}G^{\circ} = \Delta_{sol}H^{\circ} - T\Delta_{sol}S^{\circ}$

$\therefore \Delta_{sol}S^{\circ} = (\Delta_{sol}H^{\circ} - \Delta_{sol}G^{\circ})/T = (-600 \text{ J mol}^{-1} + 16800 \text{ J mol}^{-1})/$ $298.15 \text{ K} = +16200/298.15 = +54.3 \text{ J K}^{-1} \text{ mol}^{-1}$.

The values given in Table 3.9 refer to the production of aqueous solutions in which the molar concentration of the ions is 1 mol dm^{-3}, and do not bear a quantitative relationship to the solubilities of the salts. However, the thermodynamic parameters give a good insight as to whether the salts should be very soluble or otherwise. The positive value of $\Delta_{sol}G^{\circ}$ for the solution of NaF indicates that the salt is unlikely to be very soluble. Sodium chloride is very soluble, consistent with the negative value of $\Delta_{sol}G^{\circ}$, and Table 3.9 shows that it is the positive entropy change that makes the value negative. With NaBr and NaI, both enthalpic and entropic factors combine to make the $\Delta_{sol}G^{\circ}$ values negative, and both salts are even more soluble than NaCl.

Box 3.1 Relationship between Solubility and Gibbs Energy Change for Solution

If activity coefficients are ignored (assumed to be unity: a gross approximation responsible for the non-quantitative connection between changes in Gibbs energy of solution and actual salt solubilities), it is possible to draw up a table of values of the change in Gibbs energy for the solution of a compound in water that might be expected for various solubilities. Table 3.10 contains the calculations of $\Delta_{sol}G^{\ominus}$ for various solubilities of 1:1 ionic compounds. The calculations are based on the approximate relationship:

$$-\Delta_{sol}G^{\ominus} = RT\ln c^2 \qquad (3.30)$$

where c is the molar concentration of both of the ions compared to the standard molar concentration (1 mol dm^{-3}).

The figures given in Table 3.10 may be used to define a qualitative scale of solubility for compounds. This division is given in Table 3.11.

This assumes that the activity of the solid compound is unity.

Table 3.10 Gibbs energy of solution formation for 1:1 compounds

Solubility, c/mol dm^{-3}	$\Delta_{sol}G^{\ominus}$/kJ mol^{-1}
10	−11.4
1	0
1×10^{-1}	11.4
1×10^{-3}	34.2
1×10^{-6}	68.4
1×10^{-9}	102.6

With the above reservations in mind, the solubilities of all the Group 1 halides may be considered further. The overall Gibbs energy change when a salt dissolves in water may be divided into two stages: the initial destruction of the crystal lattice to produce gaseous ions, followed by the hydration of the ions to form the aqueous solution. This is shown by the thermochemical cycle in Figure 3.2. The values of $\Delta_{sol}H^{\ominus}$(MX, s) and $\Delta_{sol}S^{\ominus}$(MX, s) pertain to one mole of salt dissolving to give an aqueous solution that has molar concentrations of both hydrated ions of 1 mol dm^{-3}. The first stage is endothermic to the extent of the lattice enthalpy and is accompanied by a considerable increase in entropy as the ordered solid is transformed into the chaotic motion of the gas phase.

Table 3.11 Qualitative solubility ranges

Range of solubility/ mol dm^{-3}	Description of compound solubility
Below 1.0 × 10^{-6}	Insoluble
1.0 × 10^{-6} to 1.0 × 10^{-3}	Sparingly soluble
1.0 × 10^{-3} to 1.0	Soluble
Higher than 1.0	Very soluble

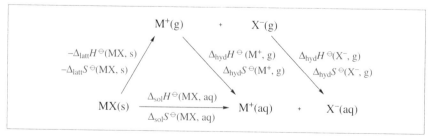

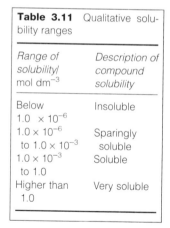

Figure 3.2 A thermochemical cycle for the formation of a solution of compound MX in water

The second stage is the hydration of the gaseous ions, and is exothermic because of the attractive forces operating between the ions and the polar water molecules. There is an accompanying reduction in entropy, as the gas phase ions have their motions constrained to a particular volume (1 dm^3) and the ions, as they are hydrated, cause the restriction of motion of a number of water molecules, leading to a further entropy reduction.

In this treatment the enthalpy changes are considered first, followed by the entropy changes, and finally the enthalpy and entropy data are combined to give the Gibbs energy changes.

The negative values of the lattice enthalpies are plotted against the cation radii in Figure 3.3. The negative values represent the enthalpy changes accompanying the conversion of the solid compounds to their gaseous constituent ions (the opposite of lattice formation).

The numerical details of the values of $\Delta_{sol}H^\ominus$, $\Delta_{sol}S^\ominus$ and $\Delta_{sol}G^\ominus$ are not given, but they are all calculated from data given in the text and the literature values for the standard entropies of the solid compounds.

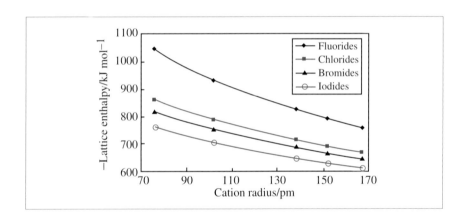

Figure 3.3 Plots of $-\Delta_{latt}H^\ominus$ against cation radius for the Group 1 halides

The trends in Figure 3.3 are best interpreted by using the using the theoretical Born–Landé equation for the enthalpy change for lattice formation from the constituent gaseous ions:

$$\Delta_{latt}H^\ominus = -\frac{N_A M z_+ z_- e^2}{4\pi\varepsilon_0 (r_+ + r_-)}\left(1 - \frac{1}{n}\right) - \frac{5}{2}(m-1)RT \qquad (3.31)$$

In equation (3.31), N_A is the Avogadro constant, M is the Madelung constant which is dependent upon the crystal structure, z_+ and z_- are the numerical charges on the ions, ε_0 is the vacuum permittivity, r_+ and r_- are the ionic radii, and n is the Born exponent (a constant dependent upon the electronic configuration of the ions). The term $\frac{5}{2}$ is included to convert internal energy change to enthalpy change, and m is the number of ions produced per formula unit. For the Group 1 halides, $m = 2$, and the term has a value of 6.2 kJ mol^{-1}, as discussed in Section 2.4.1.

The Group 1 halides have the NaCl structure (6:6 coordination) except for the chloride, bromide and iodide of caesium, which have the CsCl structure (8:8 coordination). The plots shown in Figure 3.3 show a general decrease in the negative value of $\Delta_{latt}H^\ominus$ as the cation radius increases, as would be expected from equation (3.31). Also, Figure 3.3 shows that for any set of values of $\Delta_{latt}H^\ominus$ at a particular value of the cation radius, the values of $\Delta_{latt}H^\ominus$ become less negative as the anion radius increases.

Plots of the enthalpies of hydration of the pairs of ions of the Group 1 halides against the cation radii are shown in Figure 3.4.

The values of $\Delta_{hyd}H^{\oplus}$ are the appropriate sums of the data given in Tables 2.8 (cations) and 2.12 (anions).

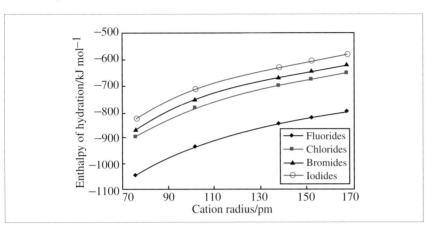

Figure 3.4   Plots of $\Delta_{hyd}H^{\oplus}$ against cation radius for the pairs of ions of the Group 1 halides

Figure 3.4 shows that the values of $\Delta_{hyd}H^{\oplus}$ become less negative as the radius of the cation increases and as the anions increase in radius. Such a variation would be expected from the Born equation (2.43), suitably modified for cations by the addition of 65 pm (the mean value for the Group 1 cations, see Table 2.9) to the ionic radius:

$$\Delta_{hyd}H^{\oplus} = -\frac{69800}{r_+ + 65} \qquad (3.32)$$

The values of ionic radius are in picometres and the resulting enthalpy change is given in units of kJ mol^{-1}. For anions the equation becomes:

$$\Delta_{hyd}H^{\oplus} = -\frac{69800}{r_-} \qquad (3.33)$$

This is consistent with the values of $\Delta_{hyd}H^{\oplus}$ becoming less negative as the anion radius increases, for any of the Group 1 halides. Note that for any particular Group 1 cation the sum of the enthalpies of hydration of the cation/fluoride combination is considerably more negative than the other three values. The cation and anion radii are separate as indicated by equations (3.32) and (3.33), and the relatively small fluoride ion has a more dominant influence than it does on lattice enthalpies, which are dependent on the sum of the cation and anion radii.

The overall enthalpy changes for the solution of the compounds, $\Delta_{sol}H^{\oplus}$ (MX, s), are derived from the equation:

$$\Delta_{sol}H^{\oplus}(MX, s) = -\Delta_{latt}H^{\oplus}(MX, s) + \Delta_{hyd}H^{\oplus}(M^+, g)$$
$$+ \Delta_{hyd}H^{\oplus}(X^-, g) \qquad (3.34)$$

The results are plotted against the cation radii in Figure 3.5.

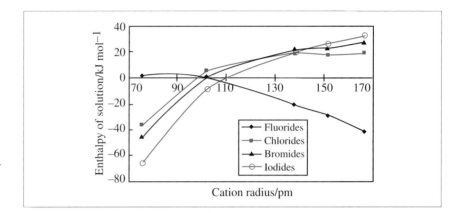

Figure 3.5 Plots of the enthalpies of solution, $\Delta_{sol}H^{\ominus}$, of the Group 1 halides against cation radius

The two sets of data, individually showing relatively smooth variations of $\Delta_{latt}H^{\ominus}$ and $\Delta_{hyd}H^{\ominus}$ with cation radii, combine to give plots in which there are still reasonably regular variations. That for the fluoride compounds is distinctly different from the plots of the other three series. The reason for this distinction is connected with the small ionic radius of the fluoride ion, as indicated above.

The enthalpy-only data shown in Figure 3.5 indicate that the compounds LiCl, LiBr and LiI, and the fluorides of K, Rb and Cs, might be expected to be very soluble, having very negative values of $\Delta_{sol}H^{\ominus}$. That the values of $\Delta_{sol}S^{\ominus}$ are also important is shown by a study of Figure 3.6. For ease of comparison of the entropy data with those for the enthalpies of solution, the former are presented as values of $T\Delta_{sol}S^{\ominus}$. The entropy data, plotted against cation radii in Figure 3.6, are obtained from the equation:

$$T\Delta_{sol}S^{\ominus}(MX, s) = -T\Delta_{latt}S^{\ominus}(MX, s) + T\Delta_{hyd}S^{\ominus}(M^+, g)$$
$$+ T\Delta_{hyd}S^{\ominus}(X^-, g) \qquad (3.35)$$

The entropies of solution are almost all positive, the positive values of the entropies of lattice vaporization predominating over the negative entropies of hydration of the pairs of ions.

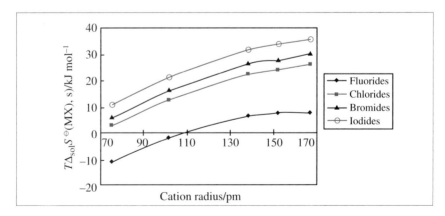

Figure 3.6 Plots of the $T\Delta_{sol}S^{\ominus}(MX, s)$ against cation radius for the pairs of ions contained by the Group 1 halides

The mainly positive entropies of solution shown in Figure 3.6 indicate that solubility is favoured by the larger ions. LiF is least favoured and CsI is most favoured.

The final stage in the presentation is carried out by using equation (3.36):

$$\Delta_{sol}G^{\ominus} = \Delta_{sol}H^{\ominus} - T\Delta_{sol}S^{\ominus} \tag{3.36}$$

The results are given in Table 3.12 for the Gibbs energies of solution of the solid Group 1 halides, and are plotted against the cation radii in Figure 3.7.

> Remember that an increase of entropy favours a process because the $T\Delta S$ term contributes a negative quantity to the value of ΔG.

Table 3.12   Gibbs energies of solution of the Group 1 halides (kJ mol^{-1})

	F	Cl	Br	I
Li	+15	−40	−56	−74
Na	+6	−9	−16	−27
K	−25	−5	−6	−10
Rb	−34	−7	−5	−5
Cs	−45	−8	−4	−1

The very negative values of $\Delta_{sol}G^{\ominus}$(MX, s) in Table 3.12 are shown in colour.

Comparison of Figures 3.5 and 3.7 indicates the effects of the $\Delta_{sol}S^{\ominus}$ values. The effects of the favourable entropy changes show that only LiF is expected to have low solubility.

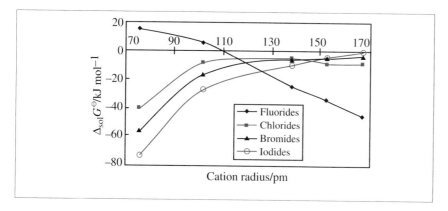

Figure 3.7 Plots of the Gibbs energies of solution against cation radius for the Group 1 halides

Experimental values of the Gibbs energies of solution of the Group 1 halides are given in Table 3.13. The entries in colour are those values that

The extent of the agreement between the calculated and experimental values for the Gibbs energies of solution gives confidence in the calculated values, and in the conclusions reached about the factors producing the variations in the values.

are more negative than -20 kJ mol^{-1}, and indicate that the particular compounds would be expected to show considerable solubility.

Table 3.13 Experimental Gibbs energies of solution, $\Delta_{sol}G^{\circ}$, of the Group 1 halides (kJ mol^{-1})

	F	Cl	Br	I
Li	15.8	−40.2	−55.3	−74.5
Na	5.6	−9.0	−16.8	−27.4
K	−24.3	−6.0	−6.5	−10.0
Rb	−37[a]	−7.4	−6.1	−6.7
Cs	−45.3	−8.7	−4.6	−3.0

[a] Estimated.

Table 3.14 gives the solubilities of the Group 1 halides at 25 °C in terms of the molalities of their saturated solutions. Molalities of solutes of 12.0 and greater are indicated in colour.

Table 3.14 Solubilities of the Group 1 halides at 25 °C expressed as molalities

	F	Cl	Br	I
Li	0.052	20.42	20.84	12.34
Na	0.98	6.15	9.27	12.27
K	17.53	4.77	5.74	8.92
Rb	28.7	7.76	7.08	7.78
Cs	37.7	11.34	5.81	3.26

Figure 3.8 is a plot of the solubilities of the Group 1 halides as molalities of the compounds in their saturated solutions at 25 °C against their cation radii. The general form of the plots is roughly the inverse of that of the plots in Figure 3.7, as would be expected from equation (3.30).

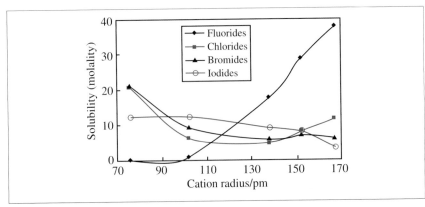

Figure 3.8 Plots of the solubilities of the Group 1 halides as molalities of the compounds in their saturated solutions at 25 °C against cation radius

The solubilities of the Group 1 halides are fairly consistent with the thermodynamic data of Table 3.13. The values of $\Delta_{sol}G^{\ominus}$ that are more negative than -20 kJ mol^{-1} are shown in colour, as are the solubilities of the very soluble halides. Lithium and sodium fluorides are not very soluble and CsI is the least soluble caesium salt and the least soluble Group 1 halide. The pattern of solubilities in Table 3.14 is roughly the inverse of that of the $\Delta_{sol}G^{\ominus}$ values of Table 3.13. The very soluble three larger halide salts are those with the small lithium cation, and the very soluble three larger Group 1 salts are those with the small fluoride anion.

The solubilities of the Group 1 halides indicate that solubility is associated with those halides in which there are large differences in size between the cations and anions. When the differences are small, the solubilities are relatively low. As is shown in Chapter 2, the cations behave in general in solution as though they possessed radii that are larger by 65 pm than their respective ionic radii. Table 3.15 gives the appropriate cation–anion radii differences to which 65 pm have been added.

Table 3.15 Differences between the radii of cations to which 65 pm have been added and the radii of anions; values in pm

	F	Cl	Br	I
Li	+8	−40	−55	−79
Na	+34	−4	−29	−53
K	+70	+22	+7	−17
Rb	+84	+36	+21	−3
Cs	+99	+51	+36	+12

The values in Table 3.15 shown in colour, corresponding to large values of the adjusted cation–anion differences (those outside the range ± 40 pm), are for the compounds that are very soluble in water, and should be compared with the solubilities given in Table 3.14. An almost identical pattern is observed for the values of the differences between the enthalpies of hydration of the cations and anions, given in Table 3.16. The entries in colour are outside the range ± 100 pm.

Table 3.16 Differences between the enthalpies of hydration of the cations and anions; values in kJ mol^{-1}

	F	Cl	Br	I
Li	−34	−179	−210	−251
Na	+80	−65	−96	−137
K	+164	+19	−12	−53
Rb	+189	+44	+13	−28
Cs	+213	+68	+37	−4

Again the very soluble halides have their enthalpy differences in colour, and the pattern of coloured entries mirrors that of the solubilities in Table 3.14. A correlation between $\Delta_{sol}H^{\ominus}$ and the difference $[\Delta_{hyd}H^{\ominus}(X^-, g) - \Delta_{hyd}H^{\ominus}(M^+, g)]$ was noted by Fajans[3] in 1921. A similar correlation using Gibbs energy values is more logical and Figure 3.9 shows a plot of $\Delta_{sol}G^{\ominus}$ for the Group 1 halides against the difference $[\Delta_{hyd}G^{\ominus}(X^-, g) - \Delta_{hyd}G^{\ominus}(M^+, g)]$. The two points that are well above the trend line are for NaF on the left and LiF on the right of the plot, both compounds showing relatively poor solubility.

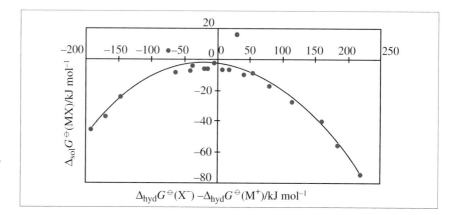

Figure 3.9 A plot of $\Delta_{sol}G^{\ominus}$ for the Group 1 halides against the difference $[\Delta_{hyd}G^{\ominus}(X^-, g) -\Delta_{hyd}G^{\ominus}(M^+, g)]$

The trend line shows that there is indeed the correlation suggested by Fajans. Large values, positive or negative, of $[\Delta_{hyd}G^{\ominus}(X^-, g) -\Delta_{hyd}G^{\ominus}(M^+, g)]$ are associated with considerably negative values of $\Delta_{sol}G^{\ominus}(MX, s)$ and would therefore be expected to be the more soluble. It bears out the rule he suggested (for enthalpy changes) that with different salts with a common cation (or common anion) the solubility will be a minimum when $\Delta_{hyd}G^{\ominus}(X^-, g) = \Delta_{hyd}G^{\ominus}(M^+, g)$. This suggests that the connection between $[\Delta_{hyd}G^{\ominus}(X^-, g) - \Delta_{hyd}G^{\ominus}(M^+, g)]$ for any of the compounds and the extent to which they separate when the crystal lattice comes into contact with water is related to the extent to which one ion (the smaller one) interacts more strongly with the molecules of water than the other (the larger one). Such an imbalance of ion–water interactions is energetically more advantageous than the case where two equally sized ions compete with the same attractive force for water molecules to complete their hydration spheres. In the case of a very soluble compound there are very few water molecules per ion and the larger ion might not fulfil its expected quota.

3.6.1 Solubilities of Salts containing Doubly Charged Ions

In this section the solubilities of NaCl, $CaCl_2$, Na_2CO_3 and $CaCO_3$ are compared to investigate the effects of having doubly charged ions in a compound. The solubilities of the compounds are given in Table 3.17 in terms of mass % and molar concentration. Sodium and calcium cations are chosen for this comparison because they have almost identical ionic radii: Na^+ 102 pm, Ca^{2+} 100 pm.

Table 3.17 Solubility data for sodium and calcium chlorides, carbonates and sulfates

Salt	Solubility/ mass%	Density/ kg m^{-3}	Mass of water in 1 dm^3 of solution/g	Mass of salt in 1 dm^3 of solution/g	Molar concentration of salt/mol dm^{-3}
NaCl	26.39	1200	833	317	5.42
$CaCl_2$	44.83	1460	805	655	5.0
Na_2CO_3	17.92	1189	976	213	2.01
$CaCO_3$				5.8×10^{-3}	5.8×10^{-5}
Na_2SO_4	21.94	1219	952	267	1.88
$CaSO_4$	0.21			2.1	0.015

The two chlorides have almost identical solubilities, except that the NaCl solution has a molar concentration of ions of $2 \times 5.4 = 10.8$ mol dm^{-3} and that of the $CaCl_2$ solution is $3 \times 5.0 = 15.0$ mol dm^{-3}. Sodium carbonate solution (molar ion concentration $= 6.03$ mol dm^{-3}) is less concentrated than that of sodium chloride (molar ion concentration $= 10.8$ mol dm^{-3}). When both cations and anions are doubly charged, as in $CaCO_3$, the relatively high lattice enthalpy predominates to make the compound insoluble. Calcium sulfate is only sparingly soluble, but sodium sulfate is soluble. The double charge on the calcium ion favours a larger lattice enthalpy compared to the enthalpy of hydration.

Most minerals contain metals with oxidation states equal to or greater than two and coupled with doubly charged anions, *e.g.* oxide or sulfide, are insoluble in water and consequently persist in the Earth's surface layer. Naturally occurring inorganic compounds consisting of metal ions in the +1 or +2 states and with counter ions such as halides are soluble in water, and are continually leached into the oceans (see Table 1.1). In arid parts of the world they occur as evaporite deposits, *e.g.* the salt flats near Salt Lake City. Evaporite salts are precipitated from concentrated brines (concentrated natural waters) in the reverse order of their solubilities under local conditions: $CaCO_3$ (limestone), $CaSO_4.2H_2O$ [gypsum, but partially converted to anhydrite ($CaSO_4$) by increased

The vast majority of limestone deposits have their origin as remains of the shells of marine organisms. Chalk is the direct remains of shells of marine organisms and has the aragonite structure. Shells that have redissolved in the oceans (at lower depths where the pH is lower) are the source of the calcium in limestone that is formed by crystallization in the calcite structure.

temperature (above $56°C$) and pressure after deposition], NaCl, $MgSO_4$ hydrates, KCl and $MgCl_2.6H_2O$.

3.6.2 Properties that Favour Solubility of Ionic Compounds

The rules given in this section apply to the main group cations and further consideration is required to take in the properties of the salts of the transition elements. For example, $ZnCl_2$ is around 55% more soluble than $MgCl_2$, even though the cations are of almost identical ionic radius: Zn^{2+} 74 pm, Mg^{2+} 72 pm. Although the two cations have identical charge, the $3d^{10}$ filled shell does not give as efficient shielding to the Zn nucleus as that offered by the inert gas configuration of the Mg^{2+} ion. The Zn^{2+} ion therefore has a greater effective nuclear charge and in consequence a more negative enthalpy of hydration. Rule 4 should state that for common anions, transition metal salts are more soluble than those of the s- and p-block elements of similar cation radii and equal cationic charges. What has been omitted from this discussion, because of lack of space and considerable complexity, is the formation of ion pairs and ion triplets, and complex ions in general. The considerable solubility of $Al_2(SO_4)_3$ may be explained by the presence in solution of the ion pair $AlSO_4^+$ (aq) and the ion triplet $Al(SO_4)_2^-$ (aq). The ions of the transition elements may also participate in complex ion formation to enhance their solubilities. For example, $FeCl_3$ is very soluble in water because of (i) hydrolysis to give ions with charges lower than +3 such as $[Fe(H_2O)_5OH]^{2+}$, and (ii) complex ion formation to give ions such as $[Fe(H_2O)_5Cl]^{2+}$. The general insolubility of hydroxides of metals in oxidation states greater than +1 is discussed in Chapter 7.

A summary of the discussion of solubility in this section is presented as properties of ions that *favour* solubility. The general "rules" are only guidelines, since there are exceptions and given with suitable examples.

1. Single charges on cation and anion.
2. Large differences in cation and anion radii, mirrored by large differences in their enthalpies and Gibbs energies of hydration.
3. For other than single charges on either ion, that of the counter ion should be single; if both ions have charges greater than one, the solubility tends to be low.

***

Box 3.2   Examples of Solubilities

1. All fluorides with singly charged cations are soluble in water. LiF comes into the sparingly soluble class.

2. Of the Group 1 halides, the most soluble are CsF and LiI. LiF is the least soluble of the compounds and CsI is the least soluble iodide.

3. The Group 1 carbonates and sulfates are very soluble, but those of Group 2 are mainly in the sparingly soluble/insoluble classes apart from the quite soluble Be and Mg sulfates. The Group 2 halides are all very soluble except for the fluorides; their +2 ions are relatively small and solubility is favoured by the three larger halide ions. The phosphates of sodium and potassium are quite soluble, but those of the Group 2 metals are insoluble, as is aluminium phosphate.

***

Summary of Key Points

1.   Brønsted–Lowry and Lewis definitions of acids and bases were introduced and discussed.

2.   Detailed discussions were given of (i) the weak acidity of HF compared to the other halogen halides, (ii) the weak acidity of HCN,

(iii) the increasing acid strength with the oxidation state of the central element in the oxoacid series: H_3BO_3, H_2CO_3 and HNO_3 of the second period of the Periodic Table and H_3PO_4, H_2SO_4 and $HClO_4$ of the third period.

3. The pH scale was defined and discussed.

4. Factors influencing acidic and basic behaviour in aqueous solutions were discussed.

5. Forms of ions in aqueous solution and ion hydrolysis, dependent upon sizes and charges, were discussed.

6. Solubilities of ionic compounds in water were discussed, and trends explained. The effects of ionic charges and sizes were explained.

References

1. F. A. Cotton, G. Wilkinson, C. A. Murillo and M. Bochmann, *Advanced Inorganic Chemistry*, 6th edn., Wiley, New York, 1999, page 63.
2. To ensure maximum consistency the majority of the thermodynamic data come from D. R. Lide (ed.), *Handbook of Chemistry and Physics*, 83rd edn., CRC Press, Boca Raton, Florida, 2002.
3. K. Fajans, *Naturwissenschaften*, 1921, **9**, 729.

Further Reading

C. F. Baes, Jr., and R. E. Mesmer, *The Hydrolysis of Cations*, Wiley, New York, 1976.
F. A. Cotton, G. Wilkinson, C. A. Murillo and M. Bochmann, *Advanced Inorganic Chemistry*, 6th edn., Wiley, New York, 1999.
D. C. F. Morris, *Ionic Radii and Enthalpies of Hydration of Ions*, in *Struct. Bonding*, 1968, **4**, 63, and the appendix in *Struct. Bonding*, 1969, **6**, 157, in which the use of the correlation between $\Delta_{sol}G^{\circ}(MX, s)$ and $[\Delta_{hyd}G^{\circ}(X^-, g) - \Delta_{hyd}G^{\circ}(M^+, g)]$ was first discussed.
P. W. Atkins, *Physical Chemistry*, 6th edn., Oxford University Press, Oxford, 1998.
O. Söhnel and P. Novotný, *Densities of Aqueous Solutions of Inorganic Substances*, Elsevier, Amsterdam, 1985.

Problems

3.1. The solubilities of some compounds are given below in terms of mass%. Calculate their molar concentrations. Comment on the result.

Salt	Solubility/mass%	Density/kg m^{-3}
CsCl	65.73	1925
BaCl$_2$	26.98	1292
BaSO$_4$	0.00024	1000

3.2. The table below gives solubilities for neutral *compounds* containing various cation/anion combinations as molalities at 25 °C.

	Cl^-	SO_4^{2-}	CO_3^{2-}
Na$^+$	6.1	1.98	2.9
K$^+$	4.8	0.69	8.07
Mg^{2+}	5.9	2.97	0.21
Ca^{2+}	6.2	0.015	6×10^{-4}

Predict, with explanations, the consequences of mixing equal volumes of the following solutions:
(a) Na$_2$SO$_4$ (1 molal) + KCl (2 molal)
(b) Na$_2$CO$_3$ (1 molal) + MgSO$_4$ (1 molal)
(c) K$_2$CO$_3$ (1 molal) + CaCl$_2$ (1 molal)
(d) MgCl$_2$ (1 molal) + K$_2$CO$_3$ (1 molal)
(e) CaCl$_2$ (1 molal) + Na$_2$SO$_4$ (1 molal)
(f) K$_2$CO$_3$ (1 molal) + MgSO$_4$ (1 molal)

3.3. NaClO$_4$ and KClO$_4$ have solubilities at 25 °C of 67.2 and 2.04 mass%. Account for this difference.

3.4. A saturated solution of CsF is 37.4 molal. Calculate the number of water molecules available to each ion, assuming an equal distribution between cations and anions. Comment on the result.

3.5. Construct a scheme of hydrolysis processes that culminate in the production of solid hydroxides or oxides for a hypothetical metal M in its +4 oxidation state.

4

Thermodynamics and Electrode Potentials

This chapter is devoted to the important relationship between **electrode potentials** and the **changes in Gibbs energy (ΔG)** for half-reactions and overall reactions. In discussions of the properties of ions in aqueous solution it is frequently more convenient to represent changes in Gibbs energy, quoted with units of kJ mol^{-1}, in terms of electrode potentials, quoted with units of volts (V). The electrochemical series is introduced. The properties of the hydrated electron are described.

Aims

At the end of this chapter you should understand:

- The relationship between Gibbs energy changes and reduction potentials
- How to calculate the overall reaction potential from two half-reaction reduction potentials
- How to calculate the reduction potential of a half-reaction from two other half-reaction potentials
- How to construct a fully balanced equation from two half-reactions
- The criteria for reaction feasibility and for the oxidizing or reducing powers of the various oxidation states of an element
- The reducing power of the hydrated electron

4.1 Changes in Standard Gibbs Energy and their Relation to Standard Electrode Potentials

Standard electrode potentials for half-reactions and overall reactions are quoted with units of volts (V). However many electrons are transferred in the reaction or half-reaction, the potentials always apply to **one-electron changes**. The changes in standard Gibbs energy (ΔG°) for the same processes are given in kJ mol^{-1} and take into account the **total changes** in oxidation states of the reactants, *i.e.* the actual number of electrons transferred in the reaction.

In any discussion of the thermodynamic stability and **reduction–oxidation (redox)** properties of ions the fundamental equation used is:

$$\Delta G^\circ = -RT \ln K \qquad (4.1)$$

The "state" of the electrons is not defined: this is deliberate and immaterial, because the electrons cancel out when the two half-reactions are summed to give an overall balanced equation. This also applies to any overall reaction in which the two half-reactions have multipliers that ensure that the numbers of electrons associated with the reduction and oxidation processes are identical. Examples are dealt with later in the section.

which relates the value of ΔG°, the change in standard Gibbs energy, to that of the equilibrium constant, K, for any process. It is convenient to express the change in standard Gibbs energy, for any reaction, as the difference between the two standard reduction potentials for the two "half-reactions" which, when added together, produce the overall equation. For example, the overall equation:

$$Cu^{2+}(aq) + Zn(s) \rightleftharpoons Cu(s) + Zn^{2+}(aq) \qquad (4.2)$$

may be considered to be the sum of the two half-reactions:

$$Cu^{2+}(aq) + 2e^- \rightleftharpoons Cu(s) \quad \text{(reduction)} \qquad (4.3)$$

$$Zn(s) \rightleftharpoons Zn^{2+}(aq) + 2e^- \quad \text{(oxidation)} \qquad (4.4)$$

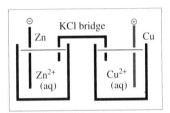

Figure 4.1   A Daniell cell

John Frederick Daniell, first professor of chemistry at King's College, London, constructed the first Cu/Zn cell in 1836. The original cell consisted of a copper vessel (the positive electrode) containing a porous clay pot. The pot was used to separate solutions of copper sulfate and zinc sulfate, the latter being inside the pot. The negative electrode was a zinc rod in the zinc sulfate solution.

The choice of potassium chloride as the "bridging" electrolyte is because the potassium and chloride ions have practically identical mobilities and a KCl bridge therefore does not produce a difference in what are known as **junction potentials** that would complicate the interpretation of the overall voltage.

Reaction (4.2) may be carried out by adding metallic zinc to a solution of copper sulfate or by using an **electrochemical cell** in which the reactants are not in contact. One such cell (a modern form of a Daniell cell) is shown in Figure 4.1.

It consists of two beakers in which the two half-reactions can occur, one containing copper(II) sulfate solution and a copper rod as the positive electrode, the other containing zinc sulfate solution with a zinc rod as the negative electrode. The two solutions are brought into electrical contact with a salt bridge – a tube containing an agar gel and potassium chloride.

When the two electrodes are connected via a high-resistance voltmeter (so that virtually no current is drawn from the cell), the measured voltage (the **electrode potential**, E°) is $+1.1$ V. When a metal wire is used to connect the two electrodes, electrons flow from the negative zinc electrode to the positive copper electrode. Zn^{2+} ions are produced in the solution and Cu^{2+} ions are reduced. The reaction would only take place within the cell if the reducing agent (Zn metal) were allowed to come into contact with the oxidizing agent (Cu^{2+} ions); the salt bridge prevents this occurrence, as the porous pot does in the original Daniell cell.

Box 4.1 The Relationship between the Standard Electrode Potential, E°, and the Change in Standard Gibbs Energy Change, $\Delta_r G^\circ$, for Any Reaction

Formal thermodynamics indicates that the change in Gibbs energy for a reaction occurring under standard conditions [10^2 kPa pressure (1 bar) and at constant temperature, 298.15 K (25 °C) unless otherwise indicated, and molar concentrations of solutes of unit activity], $\Delta_r G^\circ$, is equal, but opposite in sign, to the *maximum work* obtainable from the process, carried out under reversible conditions. If the system does work, the change in Gibbs energy is negative. Such maximum work may be estimated by measuring the voltage (e.m.f.) of an electrochemical cell in which the reaction may occur under standard conditions, where no current is drawn from the cell. E°, measured in volts, denotes the **standard electrode potential**. If the reaction occurring in the cell is associated with the passage of n moles of electrons from the reducing agent to the oxidizing agent, the amount of electrical charge is nF coulombs (F is the Faraday constant = 96485 coulomb mol^{-1} (C mol^{-1}), equal to the charge on one mole of electrons, 1.602×10^{-19} C $\times 6.022 \times 10^{23}$ mol^{-1}). The product, nFE°, gives the maximum work or energy derivable from the reaction, so the relationship between $\Delta_r G^\circ$ and E° is given by the equation:

$$\Delta_r G^\circ = -nFE^\circ \qquad (4.5)$$

where n is the number of electrons participating in the process.

The conventional way of describing the cell is by the symbolism:

$$^- Zn|Zn^{2+}(aq)||Cu^{2+}(aq)|Cu^+$$
$$\text{oxidation} \quad \text{reduction}$$

The | symbol represents the phase boundary between the solid electrode and the solution with which it is in contact, and the || symbol represents the salt bridge joining the two "half-cells" together, and assumes that the junction potentials are zero. The chemical reaction is also represented in the formalism; oxidation occurs at the left-hand electrode, reduction at the right-hand one. The right-hand electrode is positive, the left-hand one negative. The cell description is arranged so that the feasible chemical reaction is made obvious, *i.e.* for the Daniell cell, zinc metal is oxidized and $Cu^{2+}(aq)$ ions are reduced.

Box 4.2 The Significance of the Sign of E° for a Reaction

For a reaction to be feasible, the sign of the Gibbs energy change for a reaction, $\Delta_r G^\circ$, should be *negative*. Equation (4.5) shows that the criterion for feasibility of a reaction is that the sign of E° should be *positive*. In this chapter, + signs are included in positive values of E° and ΔG°.

The value of ΔG° for reaction (4.2) is -212.3 kJ mol^{-1}, corresponding to the observed maximum voltage of $+1.1$ V for a Daniell cell in which copper and zinc electrodes are in contact with solutions of their $+2$ sulfates in sulfuric acid solution that has a hydroxonium ion activity of 1 mol dm^{-1}.

Worked Problem 4.1

Q   Show that the standard electrode potential for the Daniell cell is
$+1.1$ V, given that the standard Gibbs energy change for reaction
(4.2) is -212.3 kJ mol^{-1}.

A   The reaction consists of a two-electron change: the Cu^{2+} ions
are reduced by two electrons and the Zn is oxidized by two electrons;
thus $n = 2$ in the equation $\Delta G^\circ = -nFE^\circ$, so the standard electrode
potential for the Daniell cell is given by:

$$E^\circ = -\Delta G^\circ / nF = -(-212300\,\text{J mol}^{-1}) \div (2 \times 96845\,\text{C mol}^{-1})$$
$$= +1.1\,\text{J C}^{-1} = +1.1\,\text{V}.$$

The standard electrode potential of the reaction is considered to be
equal to the *difference* between the values of the **standard reduction
potentials** of the two half-reactions, (4.3) and (4.4):

$$E^\circ = E^\circ(Cu^{2+}/Cu) - E^\circ(Zn^{2+}/Zn) \qquad (4.6)$$

where $E^\circ(Cu^{2+}/Cu)$ is the reduction potential for the half-reaction (4.3)
and $E^\circ(Zn^{2+}/Zn)$ is the *reduction* potential for the half-reaction:

$$Zn^{2+}(aq) + 2e^- \rightleftharpoons Zn(s) \qquad (4.7)$$

this being the reverse of equation (4.4).
Consider the meaning of equation (4.6) in terms of ΔG° values. The
value of ΔG° for the reduction of Cu^{2+} to Cu is:

$$\Delta G^\circ(Cu^{2+}/Cu) = -2FE^\circ(Cu^{2+}/Cu) \qquad (4.8)$$

and the value for the reduction of Zn^{2+} to Zn (the reverse of equation 4.4)
is:

$$\Delta G^\circ(Zn^{2+}/Zn) = -2FE^\circ(Zn^{2+}/Zn) \qquad (4.9)$$

the values of $n = 2$ being included because both half-reactions are two-
electron reductions.
Equation (4.2) may be constructed from its constituent half-reactions
(4.3) and (4.7), the appropriate value for its $\Delta_r G^\circ$ being the *sum* of the
ΔG° values for the half-reactions as shown:

$$Cu^{2+}(aq) + 2e^- \rightleftharpoons Cu(s) \qquad \Delta G^\circ = -2FE^\circ(Cu^{2+}/Cu)$$
$$Zn(s) \rightleftharpoons Zn^{2+}(aq) + 2e^- \qquad \Delta G^\circ = +2FE^\circ(Zn^{2+}/Zn)$$
$$Cu^{2+}(aq) + Zn(s) \rightleftharpoons Cu(s) + Zn^{2+}(aq)$$
$$\Delta_r G^\circ = -2\,FE^\circ(Cu^{2+}/Cu) + 2FE^\circ(Zn^{2+}/Zn)$$

Because the $E^\circ(Zn^{2+}/Zn)$ used in the second equation is the reduction
potential and the half-reaction as written is an oxidation, the *sign* of the

$\Delta_r G^\circ$ is reversed. $\Delta_r G^\circ$ for the reduction of Zn^{2+} to the metal is $-2FE^\circ(Zn^{2+}/Zn)$, that for the *oxidation* of Zn to Zn^{2+} is *opposite in sign*: $+2FE^\circ(Zn^{2+}/Zn)$. The standard potential for the overall reaction (4.2) is calculated by using equation (4.5):

$$E^\circ(\text{Daniell cell}) = -\Delta_r G^\circ/2F = E^\circ(Cu^{2+}/Cu) - E^\circ(Zn^{2+}/Zn) \quad (4.10)$$

which is equal to the *difference* between the two standard reduction potentials.

Since neither of the two half-reaction potentials is known absolutely, it is necessary to propose an **arbitrary zero**, relative to which all half-reaction potentials may be quoted. The half-reaction chosen to represent the arbitrary zero is the hydrogen electrode[1] in which the half-reaction is the reduction of the aqueous hydrogen ion to gaseous dihydrogen:

$$H^+(aq) + e^- \rightleftharpoons \tfrac{1}{2} H_2(g) \quad (4.11)$$

where the state of the electron is not defined. The standard reduction potential for the reaction is taken to be zero: $E^\circ(H^+/\tfrac{1}{2}H_2) = 0$. This applies strictly to standard conditions, where the activity of the hydrogen ion is unity and the pressure of the hydrogen gas is 10^2 kPa at 298.15 K.

It is conventional to quote *reduction potentials* for the general half-reaction:

$$\text{Oxidized form} + ne^- \rightleftharpoons \text{Reduced form} \quad (4.12)$$

i.e. the half-reaction written as a reduction process.

The accepted values of the standard reduction potentials for the Cu^{2+}/Cu and Zn^{2+}/Zn couples are $+0.34$ V and -0.76 V, respectively. The standard potential for the Daniell cell reaction is thus:

$$E^\circ(\text{Daniell}) = +0.34 - (-0.76) = +1.1 \text{ V}$$

The overall standard potential for the reaction:

$$2Cr^{2+}(aq) + Cu^{2+}(aq) \rightleftharpoons Cu(s) + 2Cr^{3+}(aq) \quad (4.13)$$

may be calculated from the standard reduction potentials for the constituent half-reactions:

$$Cu^{2+}(aq) + 2e^- \rightleftharpoons Cu(s) \quad E^\circ = +0.34 \text{ V} \quad (4.14)$$

$$Cr^{3+}(aq) + e^- \rightleftharpoons Cr^{2+}(aq) \quad E^\circ = -0.41 \text{ V} \quad (4.15)$$

as:

$$E^\circ(4.13) = +0.34 - (-0.41) = +0.75 \text{ V}$$

The sign of the chromium half-reaction potential is reversed because the half-reaction itself is reversed to make up the overall equation.

Attention to signs of E° and $\Delta_r G^\circ$
It is most important to give careful attention to the signs of E° and $\Delta_r G^\circ$ values in calculations. For any reaction, full or half, the signs of E° and $\Delta_r G^\circ$ are opposite to each other. If any reaction, full or half, is reversed, the signs of E° and $\Delta_r G^\circ$ must also be reversed.

Notice that the 2:1 stoichiometry of equation (4.13) has no effect upon the calculation of the overall E° value. This is because E° values represent the values of the standard Gibbs energy changes *per electron*, and since the number of electrons cancels out in an overall reaction it has no influence upon the calculation.

The overall change in the standard Gibbs energy for reaction (4.13) is given by:

$$\Delta_r G^\circ = -nFE^\circ = -2F \times 0.75 \text{ V} = -144.7 \text{ kJ mol}^{-1}$$

The *negative* value indicates that the reaction is thermodynamically feasible, as does the *positive* value for the overall potential.

It is often necessary to calculate the E° value of a half-reaction from two or more values for other half-reactions. In such cases, the stoichiometry of the overall reaction is important, and care must be taken to include the number of electrons associated with each half-reaction. For example, the two half-reactions:

$$\text{Cr}_2\text{O}_7^{2-}(\text{aq}) + 14\text{H}^+(\text{aq}) + 6\text{e}^- \rightleftharpoons 2\text{Cr}^{3+}(\text{aq}) + 7\text{H}_2\text{O}(\text{l})$$
$$E^\circ = +1.33 \text{ V} \qquad (4.16)$$

$$\text{Cr}^{3+}(\text{aq}) + \text{e}^- \rightleftharpoons \text{Cr}^{2+}(\text{aq}) \quad E^\circ = -0.41 \text{ V} \qquad (4.17)$$

contribute to the third half-reaction:

$$\text{Cr}_2\text{O}_7^{2-}(\text{aq}) + 14\text{H}^+(\text{aq}) + 8\text{e}^- \rightleftharpoons 2\text{Cr}^{2+}(\text{aq}) + 7\text{H}_2\text{O}(\text{l}) \qquad (4.18)$$

This is obtained by doubling reaction (4.17) before the addition, so that $\text{Cr}^{3+}(\text{aq})$ is eliminated. The simple addition of the two half-reaction potentials gives $1.33 - 0.41 = +0.92$ V, which is not correct for the potential of the half-reaction (4.18). Calculation of E° (4.18) via the value of $\Delta_r G^\circ$ illustrates why the correct answer is *not* the arithmetic sum of the two half-reaction potentials.

For reaction (4.16):

$$E^\circ = +1.33 \text{ V and } \Delta G^\circ = -6F \times 1.33 = -7.98F$$

For the reduction:

$$2\text{Cr}^{3+}(\text{aq}) + 2\text{e}^- \rightleftharpoons 2\text{Cr}^{2+}(\text{aq}), \quad E^\circ = -0.41 \text{ V}$$

but:

$$\Delta G^\circ = -2F \times -0.41 = 0.82F$$

because of the doubling. So, for reaction (4.18):

$$\Delta G^\circ = -7.98F + 0.82F = -7.16F$$

and:

$$E^\circ = -7.16F \div (-8F) = +0.895 \text{ V}$$

The final E° value is the sum of the individual E° values appropriately *weighted* by their respective numbers of electrons, the sum then being divided by the total number of electrons transferred in the final equation. This is because the electrons do not cancel out in such cases.

Worked Problem 4.2

Q The standard reduction potentials of the half-reactions for the reduction of diiodine to iodide ion and for the reduction of I^V to diiodine are $+0.535$ V and $+1.2$ V, respectively. Write down the fully balanced equations for the two half-reactions, and calculate the standard reduction potential for the half-reaction in which I^V is reduced to iodide ion.

A The five-electron reduction of I^V to elemental diiodine may be written as: $I^V(aq) + 5e^- \rightleftharpoons \frac{1}{2} I_2(s)$. To balance this half-reaction requires three steps. The first is to include the full formula for the I^V as IO_3^-:

$$IO_3^- (aq) + 5e^- \rightleftharpoons \frac{1}{2}I_2(s)$$

The second step is to balance the oxygen atoms by including sufficient water molecules on the right-hand side, three in this case:

$$IO_3^- (aq) + 5e^- \rightleftharpoons \frac{1}{2}I_2(s) + 3H_2O(l)$$

The final step is to balance the hydrogen atoms and the charges by including six protons (in this case) on the left-hand side:

$$IO_3^- (aq) + 6H^+(aq) + 5e^- \rightleftharpoons \frac{1}{2}I_2(s) + 3H_2O(l)$$

The two half-reactions are:

$$\frac{1}{2}I_2(s) + e^- \rightleftharpoons I^-(aq) \qquad E^\circ = +0.535 \text{ V}$$

$$IO_3^- (aq) + 6H^+(aq) + 5e^- \rightleftharpoons \frac{1}{2}I_2(s) + 3H_2O(l) \qquad E^\circ = +1.2 \text{ V}$$

Their sum gives the third half-reaction:

$$IO_3^- (aq) + 6H^+(aq) + 6e^- \rightleftharpoons 3H_2O(l) + I^-(aq)$$

which has a reduction potential:

$$E^\circ = [0.535 + (1.2 \times 5)] \div 6 = +1.09 \text{ V}$$

[*Not* the arithmetic mean of the two potentials.]

Worked Problem 4.3

Q The reduction potentials of the couples of the oxidation states of the metal M are M^{3+}/M^{2+}, $+1.0$ V, and M^{2+}/M, -1.66 V. Calculate the reduction potential for the M^{3+}/M couple.

A The two half-reactions given are:

$$M^{3+}(aq) + e^- \rightleftharpoons M^{2+}(aq) \qquad E^\circ = +1.0 \text{ V}$$

$$M^{2+}(aq) + 2e^- \rightleftharpoons M(s) \qquad E^\circ = -1.66 \text{ V}$$

Adding gives the third half-reaction:

$$M^{3+}(aq) + 3e^- \rightleftharpoons M(s)$$

for which:

$$E^\circ = [+1.0 + (2 \times -1.66)] \div 3 = -0.77 \text{ V}$$

On the hydrogen scale, the range of values for standard reduction potentials varies from about $+3.0$ V to -3.0 V for solutions in which the activity of the hydrated hydrogen ion is 1. It is also conventional to quote values for standard reduction potentials in basic solutions with a unit activity of hydrated hydroxide ion, $E_B{}^\circ$. Typical values of $E_B{}^\circ$ range between about $+2.0$ V to -3.0 V.

Examples of half-reactions at each extreme of the E° and $E_B{}^\circ$ ranges are:

Many half-reaction potentials are calculated from thermochemical data and are not relevant to any electrochemical cell measurements. This is particularly the case where solids other than metals appear in half-reactions.

$\tfrac{1}{2}F_2(g) + e^- \rightleftharpoons F^-(aq)$	$E^\circ = +2.87$ V
$Li^+(aq) + e^- \rightleftharpoons Li(s)$	$E^\circ = -3.04$ V
$O_3(g) + H_2O(l) + 2e^- \rightleftharpoons O_2 + 2OH^-(aq)$	$E_B^\circ = +1.23$ V
$Ca(OH)_2(s) + 2e^- \rightleftharpoons Ca(s) + 2OH^-(aq)$	$E_B^\circ = -3.03$ V

A listing of half-reaction reduction processes in decreasing order of their E° values is called the **electrochemical series**. Extracts from the electrochemical series are given in Table 4.1 for some selected metals and in Table 4.2 for some selected non-metals to illustrate further the range of reduction potential values.

Table 4.1 An extract of the electrochemical series for selected metals (to two decimal places); states are omitted for brevity

Couple	$E°$/V	Couple	$E°$/V	Couple	$E°$/V	Couple	$E°$/V
Au^+/Au	+1.69	Fe^{3+}/Fe	−0.04	Zn^{2+}/Zn	−0.76	Pu^{3+}/Pu	−2.03
Pt^{2+}/Pt	+1.18	Pb^{2+}/Pb	−0.13	V^{3+}/V	−0.87	Cm^{3+}/Cm	−2.04
Ir^{3+}/Ir	+1.15	Sn^{2+}/Sn	−0.14	Cr^{2+}/Cr	−0.91	Ac^{3+}/Ac	−2.20
Pd^{2+}/Pd	+0.95	Ni^{2+}/Ni	−0.26	V^{2+}/V	−1.18	Lu^{3+}/Lu	−2.28
Hg^{2+}/Hg	+0.85	Mn^{3+}/Mn	−0.28	Mn^{2+}/Mn	−1.19	Mg^{2+}/Mg	−2.37
Ag^+/Ag	+0.80	Co^{2+}/Co	−0.28	Ti^{3+}/Ti	−1.37	La^{3+}/La	−2.38
Rh^{3+}/Rh	+0.76	In^{3+}/In	−0.34	Al^{3+}/Al	−1.66	Na^+/Na	−2.71
Tl^{3+}/Tl	+0.74	Cd^{2+}/Cd	−0.40	U^{3+}/U	−1.80	Ca^{2+}/Ca	−2.87
Cu^+/Cu	+0.52	Fe^{2+}/Fe	−0.45	Np^{3+}/Np	−1.86	Ba^{2+}/Ba	−2.91
Co^{3+}/Co	+0.46	Ga^{3+}/Ga	−0.55	Am^{3+}/Am	−1.90	Cs^+/Cs	−3.03
Cu^{2+}/Cu	+0.34	Cr^{3+}/Cr	−0.74	Lr^{3+}/Lr	−1.96	Li^+/Li	−3.04

The entries in Table 4.1 suggest several questions. Why are the reduction potentials of the Groups 1 and 2 elements so negative? What are the reasons for the trend in the reduction potentials of the Al^{3+}/Al, Mg^{2+}/Mg and Na^+/Na couples? Why do the reduction potentials of the transition metal couples vary so much? Why are the reduction potentials of the Group 11 metals (the coinage metals of yesteryear) positive? And so on. All these and other questions raised by the entries in Table 4.2 are answered in Chapters 6, 7 and 8.

Box 4.3 Interpretation of $E°$ values

The following generalizations are helpful in the interpretation of $E°$ values.

1. A highly negative value of $E°$ implies that the reduced form of the couple is a good reducing agent. For example, any Group 1 metal is a good reducing agent, its reduction potential (*e.g.* −2.71 V for Na) indicating that, written as an oxidation half-reaction:

$$Na(s) \rightleftharpoons Na^+(aq) + e^-$$

it contributes an amount $-2.71F$ to the Gibbs energy change for the overall reaction.

2. A large positive value of $E°$ implies that the oxidized form of the couple is a good oxidizing agent. For example, the reduction potential for the reduction of dichlorine to aqueous chloride ion is + 1.36 V, and the reduction half-reaction:

$$\tfrac{1}{2}Cl_2(g) + e^- \rightleftharpoons Cl^-(aq)$$

contributes an amount $-1.36F$ to the Gibbs energy change for the overall reaction.

3. The oxidized form of a couple will oxidize the reduced form of a second couple if the $E°$ value of the first couple is more positive than that of the second couple. Taking for example the two half-reactions:

$$H_2O_2(aq) + 2H^+(aq) + 2e^- \rightleftharpoons 2H_2O(l) \qquad E° = +1.76 \text{ V}$$
$$Fe^{3+}(aq) + e^- \rightleftharpoons Fe^{2+}(aq) \qquad E° = +0.77 \text{ V}$$

Table 4.2 An extract of the electrochemical series for selected non-metals (to two decimal places); states are omitted for brevity

Couple	$E°$/V
$\tfrac{1}{2}F_2/F^-$	+2.87
O_3/O_2	+2.08
$HOCl/\tfrac{1}{2}Cl_2$	+1.61
ClO_4^-/Cl^-	+1.39
$\tfrac{1}{2}Cl_2/Cl^-$	+1.36
$O_2/2H_2O$	+1.23
NO_3^-/HNO_2	+0.93
BrO_3^-/Br^-	+0.61
SO_4^{2-}/H_2SO_3	+0.17
H_3PO_3/P	−0.45
S/S^{2-}	−0.48
H_3BO_3/B	−0.87

A solution containing $Fe^{2+}(aq)$ and H_2O_2 is known as Fenton's reagent. It is used in the oxidation of organic compounds in aqueous solution. It operates via the initial transfer of one electron from the $Fe^{2+}(aq)$ ion to the H_2O_2:

$Fe^{2+}(aq) + H_2O_2$
$\rightarrow Fe^{2+}(aq) + OH^-(aq) + \cdot OH$

producing a hydroxyl free radical which may oxidize the organic substrate, RH:

$\cdot OH + RH \rightarrow R\cdot + H_2O$

Then:
$2R\cdot \rightarrow R_2$

or the radical R· **dismutes**, *e.g.*:

$2C_2H_5\cdot \rightarrow C_2H_6 + C_2H_4$

Hydrogen peroxide will oxidize iron(II) according to the overall reaction:

$$H_2O_2(aq) + 2H^+(aq) + 2Fe^{2+}(aq) \rightleftharpoons 2Fe^{3+}(aq) + 2H_2O(l)$$

The overall E° value is $+1.76 - 0.77 = +0.99$ V, which, being positive, makes the Gibbs energy change negative and the reaction thermodynamically feasible.

A usual caution here is that, although a reaction is thermodynamically feasible, it might be slow for kinetic reasons. The $H_2O_2(aq)/Fe^{2+}(aq)$ reaction is a rapid process unhindered by high activation energy.

4.2   The Hydrated Electron and Absolute Values of Reduction Potentials

The primary neutral products of the irradiation of liquid water, the hydroxyl free radical and the hydrated electron, also have a transient existence. The hydrated electron reacts with water to give a hydrogen atom and a hydroxide ion:

$e^-(aq) + H_2O(l) \rightarrow H(aq)$
$+ OH^-(aq)$

The H and OH radical species largely react with each other to give water:

$H(aq) + OH(aq) \rightarrow H_2O(l)$

with small amounts of dimerization:

$2H(aq) \rightarrow H_2(g)$

$2OH(aq) \rightarrow H_2O_2(aq)$

The hydrated electron, $e^-(aq)$, is a well-characterized chemical entity. If liquid water is subjected to irradiation by X- or γ-ray quanta, electrons (photoelectrons) are released from water molecules, which thereby acquire a positive charge:

$$H_2O + h\nu \rightarrow H_2O^+ + e^- \tag{4.19}$$

The photoelectron retains most of the energy of the incident photon and itself produces further ionizations. It produces many more H_2O^+/e^- pairs in losing its energy. The excess of energy possessed by the electrons so produced is used in such further ionizations until the electrons become thermalized, *i.e.* they have translational energies typical of the temperature of the bulk medium. Both types of ion become hydrated by interaction with the water solvent:

$$H_2O^+ + H_2O(l) \rightarrow H_3O^+(aq) + OH(aq) \tag{4.20}$$

$$e^- \rightarrow e^-(aq) \tag{4.21}$$

Dissolving metallic sodium in liquid ammonia produces more stable solvated electrons. Dilute solutions of sodium in liquid ammonia are blue and have an absorption spectrum very similar in distribution to that of the hydrated electron, the absorption maximum being at 1500 nm with a broad band extending into the visible region. The blue colour produced by dissolving any of the other Group 1 metals in liquid ammonia is identical to that of the sodium solution, indicating that the ammoniated electron produces the colour rather than being a property of the Group 1 element.

That the hydrated electron is a separate chemical entity has been demonstrated by the technique of **pulse radiolysis**. This consists of subjecting a sample of pure water to a very short pulse of accelerated electrons. The energetic electrons have the same effect upon water as a beam of γ-ray photons. Shortly after the pulse of electrons has interacted with the water, a short flash of radiation (ultraviolet and visible radiation from a discharge tube) is passed through the irradiated water sample at an angle of $90°$ to the direction of the pulse to detect the absorption spectra of any transient species. The spectrum of the hydrated electron consists of a broad absorption band with a maximum absorption at a wavelength of 715 nm, in the near-infrared region. The broad band shows considerable absorption in the red region of the visible spectrum, so a solution of hydrated electrons, if it lasted long enough to be viewed by eye, would be blue in colour.

Measurements of the rate of reaction of the hydrated electron with water and the reverse process have given a value for the change in Gibbs energy for the equilibrium:

$$e^-(aq) + H_2O(l) \rightleftharpoons H(aq) + OH^-(aq) \qquad (4.22)$$

of $+35.2$ kJ mol^{-1}. The change in Gibbs energy for the dehydration of the hydrated hydrogen atom is estimated to be -19.2 kJ mol^{-1}. The change in Gibbs energy for the dimerization of hydrogen atoms to give dihydrogen gas is -203.3 kJ mol^{-1} and that for the reaction of hydrated protons with hydrated hydroxide ions to give liquid water is -79.9 kJ mol^{-1}. These reactions may be combined in a Hess's law calculation, to give the change in Gibbs energy for the reaction of the hydrated electron with the hydrated proton to produce gaseous dihydrogen:

Reaction	$\Delta_r G^\circ$/kJ mol^{-1}
$e^-(aq) + H_2O(l) \rightleftharpoons H(aq) + OH^-(aq)$	$+35.2$
$H(aq) \rightleftharpoons H(g)$	-19.2
$H(g) \rightleftharpoons \frac{1}{2} H_2(g)$	-203.3
$H^+(aq) + OH^-(aq) \rightleftharpoons H_2O(l)$	-79.9
$e^-(aq) + H^+(aq) \rightleftharpoons \frac{1}{2} H_2(g)$	-267.2

The value of $\Delta_r G^\circ$ of -267.2 kJ mol^{-1} for the hydrated electron/hydrated proton reaction converts into a reduction potential of $267.2/96.485 = +2.77$ V on the conventional scale and indicates that the hydrated electron is approximately equal to sodium in its reducing power:

$$\frac{1}{2}H_2(g) \rightleftharpoons e^-(aq) + H^+(aq) \quad E^\circ = -2.77\,V \qquad (4.23)$$

$$Na^+(aq) + \frac{1}{2}H_2(g) \rightleftharpoons H^+(aq) + Na(s) \quad E^\circ = -2.71\,V \qquad (4.24)$$

A thermochemical cycle representing the reaction between the hydrated electron and the hydrated proton to give gaseous dihydrogen is shown in Figure 4.2.

The ammoniated electron, e^-(am), is much more stable than the hydrated electron, but its solutions do decompose slowly to give hydrogen atoms and amide ions, NH_2^- (am), initially.

It is likely that the hydrated electron is a reactive intermediate in the reactions of the Group 1 metals and the heavier Group 2 metals with liquid water, *i.e.* those that possess standard reduction potentials ≤ -2.77 V. Reaction (4.23) could be used to establish an absolute scale for standard reduction potentials, but there have been no moves to do this so far.

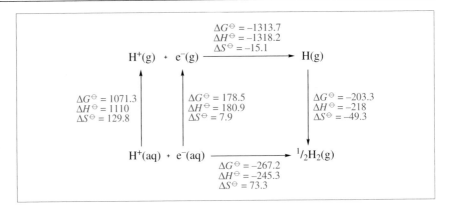

Figure 4.2 Standard Gibbs energy, standard enthalpy and standard entropy cycles for the reaction between hydrated protons and hydrated electrons to give gaseous dihydrogen; Gibbs energy and enthalpy changes are in units of kJ mol^{-1}; the entropy changes are in units of J K^{-1} mol^{-1}

Table 4.3 Standard entropies for some species

Species	S°/J K^{-1} mol^{-1}
H_2(g)	+130.7
H(g)	+114.7
H^+(g)	+108.9
H^+(aq)	−20.9
e$^-$(g)	+20.9
e$^-$(aq)	+13.0

The values of the changes in standard Gibbs energy, standard enthalpy and standard entropy are given for all the stages. The calculation of some of the values depends upon the known values for the standard entropies of the participating species given in Table 4.3.

The values in Table 4.3 for H(g), H^+(g) and e$^-$(g) are calculated by using the Sackur–Tetrode equation (2.60). Those of H(g) and e$^-$(g) contain the term $R \ln 2$ ($= 5.76$ J K^{-1} mol^{-1}) to take into account the two possible values of the electron spins. The values for the other species are taken from the chemical literature.[2,3]

Worked Problem 4.4

Q Calculate the standard molar entropy of the gaseous hydrogen atom.

A Taking into account the two possibilities for the electron spin, the standard molar entropy is given by:

$$S^{\circ} = {}^{3}\!/_{2}R \ln 1 + 108.9 + R \ln 2 = 0 + 108.9 + 5.8$$
$$= 114.7 \text{ J K}^{-1} \text{ mol}^{-1}$$

4.2.1 The Gibbs Energy Cycle

The values for Gibbs energy changes shown in Figure 4.2 are either available from the literature or are calculated from the enthalpy and entropy changes. The change in standard Gibbs energy for the dehydration of the hydrated proton:

$$H^+(aq) \rightarrow H^+(g) \qquad (4.25)$$

is calculated from the change in enthalpy (1110 kJ mol^{-1}) and the change in entropy: $+108.9 - (-20.9) = +129.8$ J K^{-1} mol^{-1}.

Worked Problem 4.5

Q Calculate the change in standard Gibbs energy for the dehydration of the hydrated proton.

A $\Delta G^{\circ}(H^+(aq) \rightarrow H^+(g)) = 1110 - (298.15 \times 129.8/1000)$
$$= 1071.3 \text{ kJ mol}^{-1}$$

The change in standard Gibbs energy for the reaction:

$$H^+(g) + e^-(g) \rightarrow H(g) \qquad (4.26)$$

is calculated from the enthalpy change of -1318.2 kJ mol^{-1} and the corresponding entropy change of $+114.7 - 108.9 - 20.9 = -15.1$ J K^{-1} mol^{-1}:

$$\Delta G^{\circ}(H^+(g) + e^-(g) \rightarrow H(g)) = -1318.2 + (298.15 \times 15.1/1000)$$
$$= -1313.7 \text{ kJ mol}^{-1}$$

The change in standard Gibbs energy for the reaction:

$$H(g) \rightarrow \frac{1}{2}H_2(g) \qquad (4.27)$$

is calculated from the enthalpy change of -218 kJ mol^{-1} and the corresponding entropy change of $+65.4 - 114.7 = -49.3$ J K^{-1} mol^{-1}:

$$\Delta G^{\circ}(H(g) \rightarrow \frac{1}{2}H_2(g)) = -218 + (298.15 \times 49.3/1000)$$
$$= -203.2 \text{ kJ mol}^{-1}$$

From the calculated values of the Gibbs energy changes, that for the dehydration of the hydrated electron may be estimated as:

$$\Delta G^{\circ}(e^-(aq) \rightarrow e^-(g)) = -267.2 - 1071.3 + 1313.7 + 203.2$$
$$= +178.5 \text{ kJ mol}^{-1}$$

4.2.2 The Enthalpy Cycle

The change in standard Gibbs energy for the reaction:

$$e^-(aq) + H^+(aq) \rightleftharpoons \frac{1}{2}H_2(g) \qquad (4.28)$$

is -267.2 kJ mol^{-1}, and the change in standard entropy, calculated from the values given in Table 4.3, is $+73.4$ J K^{-1} mol^{-1}. This makes the standard enthalpy change for the reaction equal to $-267.2 + (298.15 \times 73.4/100) = -245.3$ kJ mol^{-1}. The enthalpy changes for the other stages in the cycle of Figure 4.3 are known, and may be combined with the enthalpy change for the reaction:

$$\frac{1}{2}H_2(g) \rightleftharpoons e^-(g) + H^+(aq) \qquad (4.29)$$

to give a value for the change of enthalpy for the dehydration of the hydrated electron:

$$\Delta H^{\circ}(e^-(aq) \rightarrow e^-(g)) = -245.3 - 1110 + 1318.2 + 218$$
$$= +181 \text{ kJ mol}^{-1}$$

4.2.3 The Enthalpy of Hydration of the Electron

The enthalpy of hydration of the electron is -181 kJ mol^{-1} and corresponds to that of a relatively large ion. The value for the iodide ion (ionic radius $= 220$ pm) is -289 kJ mol^{-1}. If the simple inverse relationship between radii and the enthalpies of hydration of ions [the Born equations (2.46)] is used, the radius of the hydrated electron is estimated to be: $220 \times (-289/-181) = 362$ pm. This is a large value for an ion of single charge, and corresponds to the delocalization of the charge over a considerable volume consisting of several water molecules. This is to be expected, since the water molecule has no low-lying orbitals to accommodate an extra electron. The near-infrared spectrum of the hydrated electron has a maximum absorption at 715 nm. The quantum energy equivalent of such a wavelength is given by:

$$E = N_A h v = N_A h c / \lambda \tag{4.30}$$

$$E = [6.022 \times 10^{23}(\text{mol}^{-1}) \times 6.626 \times 10^{-34}(\text{J s}) \times 299792458(\text{m s}^{-1})]$$
$$\div [715 \times 10^{-9}(\text{m}) \times 10^3] = 167.3 \text{ kJ mol}^{-1}.$$

The quantum energy is very similar to the enthalpy of dehydration of the hydrated electron, and the absorption of the 715 nm radiation by the hydrated electron probably causes its dehydration, *i.e.* its reversion to the gaseous state, $e^-(g)$.

Worked Problem 4.6

Q Calculate the value of E° for the Na$^+$/Na reduction half-reaction using $e^-(aq)$ as the standard state of the electron.

A The conventional half-reaction for the Na$^+$/Na reduction may be written as:

$$\text{Na}^+(aq) + e^-(g) \rightleftharpoons \text{Na}(s) \qquad E^{\circ} = -2.71 \text{ V}$$

The conversion of the hydrated electron into the gaseous state:

$$e^-(aq) \rightleftharpoons e^-(g)$$

has an associated $\Delta G^\circ = +178.5$ kJ mol^{-1} that converts to an E° value of $-178500/F = -1.85$ V. Adding the two half-reactions together to give the third half-reaction:

$$Na^+(aq) + e^-(aq) \rightleftharpoons Na(s)$$

gives the associated value for $E^\circ = -2.71 - 1.85 = -4.56$ V. The very negative value is consistent with the known difficulty of reducing the $Na^+(aq)$ ion and emphasizes the reducing power of sodium metal.

Summary of Key Points

1. The relationship between Gibbs energy change and reduction potential was established.

2. Calculations of the reaction potential from two half-reaction reduction potentials were described.

3. Calculations of the reduction potential of a half-reaction from two other half-reaction potentials were described.

4. The criteria for reaction feasibility were outlined and the significance of the sign of a standard reaction potential was described.

5. The electrochemical series was introduced.

6. Generalizations for the interpretation of E° values were given.

7. The properties of the hydrated electron were described in detail.

References

1. P. W. Atkins, *Physical Chemistry*, 6th edn., Oxford University Press, Oxford, 1998, p. 257, for details of the hydrogen electrode.
2. F. S. Dainton, *Chem. Soc. Rev.*, 1975, **4**, 323.
3. M. R. C. Symons, *Chem. Soc. Rev.*, 1976, **5**, 337.

Further Reading

P. W. Atkins, *Physical Chemistry*, 6th edn., Oxford University Press, Oxford, 1998.

Problems

4.1. Write fully balanced equations for the following reactions and, using the data from Table 4.1, calculate their overall reaction potentials and indicate whether they are thermodynamically feasible or otherwise: (a) $Pt^{2+} + Sn$; (b) $Co^{2+} + V$; (c) $Mg^{2+} + Pb$; (d) $Ba^{2+} + Cd$.

4.2. Using the data of Table 4.1, calculate a value for the standard reduction potential for each of the following half-reactions: (a) $Mn^{3+}(aq) + e^{-} \rightleftharpoons Mn^{2+}(aq)$; (b) $V^{3+}(aq) + e^{-} \rightleftharpoons V^{2+}(aq)$; (c) $Fe^{3+}(aq) + e^{-} \rightleftharpoons Fe^{2+}(aq)$.

4.3. Write a fully balanced equation for the half-reaction for the reduction of $ClO_4^{-}(aq)$ to HOCl, and by using the data from Table 4.2, calculate the value of its standard reduction potential.

4.4. Calculate the value for the standard reduction potential of the half-reaction on the $e^{-}(aq)$ scale:

$$H_2O_2(aq) + 2H^{+}(aq) + 2e^{-}(aq) \rightleftharpoons 2H_2O(l)$$

given that the conventional value is $+1.78$ V.

5
The Stabilities of Ions in Aqueous Solution

This chapter is not concerned with the thermodynamic stability of ions with respect to their formation. Rather, it is concerned with whether or not a given ion is capable of existing in aqueous solution without reacting with the solvent. Hydrolysis reactions of ions are dealt with in Chapter 3. The only reactions discussed in this section are those in which either water is *oxidized* to dioxygen or *reduced* to dihydrogen. The Nernst equation is introduced and used to outline the criteria of ionic stability. The bases of construction and interpretation of Latimer and volt-equivalent (Frost) diagrams are described.

Aims

By the end of this chapter you should understand:

- The Nernst equation and its relevance to the dependence of the values of reduction potentials upon the pH of the solution
- That there is a limiting value for any reduction potential above which the oxidized form of the couple has the capacity to oxidize water to dioxygen
- That there is a limiting value for any reduction potential below which the reduced form of the couple has the capacity to reduce water to dihydrogen
- That there are practical limits to the stabilities of oxidizing and reducing agents
- How to interpret Latimer diagrams
- How to construct and interpret volt-equivalent (Frost) diagrams

5.1   The Limits of Stability of Ions in Aqueous Systems

Although reduction potentials may be estimated for half-reactions, there are limits for their values that correspond to both members of a couple having stability in an aqueous system with respect to reaction with water. For example, the Na^+/Na couple has a standard reduction potential of -2.71 V, but metallic sodium reduces water to dihydrogen. The reduced form of the couple (Na) is not stable in water. The standard reduction potential for the Co^{3+}/Co^{2+} couple is $+1.92$ V, but a solution of Co^{3+} slowly oxidizes water to dioxygen. In this case the oxidized form of the couple is not stable in water. The standard reduction potential for the Fe^{3+}/Fe^{2+} couple is $+0.771$ V, and neither oxidized form or reduced form react chemically with water. They are subject to hydrolysis, but are otherwise both stable in the aqueous system. The limits for the stability of both oxidized and reduced forms of a couple are pH dependent.

The limits of stability for water at different pH values may be defined by the **Nernst equations** for the values of reduction potentials for the hydrogen reference half-reaction:

$$H^+(aq) + e^- \rightleftharpoons \tfrac{1}{2}H_2(g) \tag{5.1}$$

and for the reduction of dioxygen to water:

$$O_2(g) + 4H^+(aq) + 4e^- \rightleftharpoons 2H_2O(l) \tag{5.2}$$

as the pH of the solution changes over the 0–14 pH range.

The equation:

$$\Delta G = \Delta G^\circ + RT \ln Q \tag{5.3}$$

represents the change in Gibbs energy for non-standard conditions of a system. The quotient, Q, is the product of the activities of the products divided by the product of the reactant activities, taking into account the stoichiometry of the overall reaction.

Equation (5.3) may be converted into the Nernst equation by using equation (4.5) so that:

$$E = E^\circ - \frac{RT}{nF} \ln Q \tag{5.4}$$

The Nernst equation may be applied to the hydrogen standard reference half-reaction and gives:

$$E(H^+/\tfrac{1}{2}H_2) = -\frac{RT}{F} \ln\left(\frac{1}{a_{H^+}}\right) \tag{5.5}$$

Walter Nernst was awarded the 1920 Nobel Prize for Chemistry for "work in thermochemistry".

The quotient, Q, has the same form as that of the equilibrium constant for the same reaction, except that the activities/concentrations are not those at equilibrium but those of any non-equilibrium reactant/product mixture.

considering the activity of dihydrogen to be unity and $E^{\circ}(\mathrm{H}^{+}/\frac{1}{2}\mathrm{H}_{2}) = 0$. In order to obtain an equation which relates E to pH, it is convenient to convert equation (5.5) to one in which logarithms to base 10 are used:

$$E(\mathrm{H}^{+}/\tfrac{1}{2}\mathrm{H}_{2}) = -\frac{2.3026\,RT}{F} \times \log_{10}\!\left(\frac{1}{a_{\mathrm{H}^{+}}}\right) \qquad (5.6)$$

At 298.15 K this simplifies to:

$$E(\mathrm{H}^{+}/\tfrac{1}{2}\mathrm{H}_{2}) = -0.0592 \times \log_{10}\!\left(\frac{1}{a_{\mathrm{H}^{+}}}\right) \qquad (5.7)$$

The theoretical definition of pH is:

$$\mathrm{pH} = -\log_{10} a_{\mathrm{H}^{+}} \qquad (5.8)$$

and substitution into equation (5.7) gives:

$$E(\mathrm{H}^{+}/\tfrac{1}{2}\mathrm{H}_{2}) = -0.0592\ \mathrm{pH} \qquad (5.9)$$

The half-reaction expressing the reduction of dioxygen to water in acid solution (equation 5.2) has an E° value of $+1.23$ V when the activity of the hydrogen ion is 1 mol dm^{-1}. The Nernst equation for the dioxygen/water couple is:

$$E(\mathrm{O}_2/2\mathrm{H}_2\mathrm{O}) = 1.23 - 0.0592\ \mathrm{pH} \qquad (5.10)$$

This assumes that the activities of dioxygen and water are both unity.

The conversion of logarithms from one base to another depends on the following algebra. By definition:

if $x = e^{y}$, then $y = \ln_{e} x$

$\log_{10} x = \log_{10}(e^{y}) = y\ \log_{10} e$

$y = \dfrac{1}{\log_{10} e} \times \log_{10} x$

$\ln_{e} x = \dfrac{1}{\log_{10} e} \times \log_{10} x$
$\qquad = 2.3026 \times \log_{10} x$

Worked Problem 5.1

Q   Derive equation (5.10). Take into account the 4H^{+} and the four electrons in equation (5.2).

A   Using the Nernst equation for the reduction of dioxygen to water:

$$E(\mathrm{O}_2/2\mathrm{H}_2\mathrm{O}) = E^{\circ} - \frac{2.3026\,RT}{4F} \times \log_{10}\!\left(\frac{1}{a_{\mathrm{H}^{+}}}\right)^{4}$$

$$= E^{\circ} - \frac{2.3026\,RT}{F} \times \log_{10}\!\left(\frac{1}{a_{\mathrm{H}^{+}}}\right)$$

$$= E^{\circ} - 0.0592\ \mathrm{pH}$$

$$= 1.23 - 0.0592\ \mathrm{pH}$$

Note how the "4"s for the numbers of hydrated protons and the stateless electrons cancel out. This is not always the case for some half-reactions, which therefore have a different **Nernst slope.** The Nernst slope is given by the factor $-2.3026RT/F$ multiplied by the ratio of the number of protons to the number of electrons in the half-reaction.

Worked Problem 5.2

Q Calculate an expression for the Nernst slope of the half-reaction:

$$H_5IO_6(aq) + H^+(aq) + 2e^- \rightleftharpoons IO_3^-(aq) + 3H_2O(l)$$

A Assuming that the activities of H_5IO_6 and IO_3^- are unity, the Nernst equation for the half-reaction is:

$$E = E^{\ominus} - \frac{2.3026RT}{2F} \times \log\left(\frac{1}{a_{H^+}}\right)^1 = E^{\ominus} - \frac{2.3026RT}{2F} \times pH$$

The Nernst slope is $-2.3026RT/2F$, i.e. $-2.3026T/F$ multiplied by $\frac{1}{2}$, which is the ratio of the number of protons to the number of electrons in the half-reaction.

Equations (5.8) and (5.10) are plotted in Figure 5.1, with pH values varying from 0 to 14. Theoretically, the oxidized form of any couple whose E value lies above the oxygen line should be unstable, and should oxidize water to dioxygen. Likewise, the reduced form of any couple whose E value lies below the hydrogen line should be unstable, and should reduce water to dihydrogen.

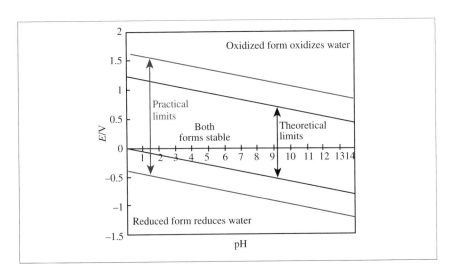

Figure 5.1 A diagram showing the theoretical and practical limits of stability of ions in solutions of pH between 0 and 14

In practice, it is found that wherever gases are evolved [in particular the necessity for initially generated (*e.g.* O or H) atoms to come together to form diatomic molecules] there is a kinetic barrier to the process, sometimes known as **overpotential.** This barrier causes the practical limits for the E values to be around 0.4 V lower than the hydrogen line or 0.4 V above the oxygen line for water to be reduced or oxidized, respectively. The practical limits are shown in Figure 5.1.

The oxidized and reduced forms of couples with E values anywhere between the two practical limits are stable in aqueous systems with respect to either the reduction or the oxidation of the solvent.

Couples with E values outside the practical limits of stability do not necessarily cause the destruction of the solvent. Some reactions, although they may be thermodynamically feasible, are kinetically very slow. The Co^{3+}/Co^{2+} couple has an E° value of $+1.92$ V, and Co^{3+} should not exist in aqueous solution. However, its oxidation of water to oxygen is very slow, and solutions containing $[Co(H_2O)_6]^{3+}$ evolve dioxygen slowly. The E° value for the Al^{3+}/Al couple is -1.66 V, and indicates that aluminium should dissolve in acidic aqueous solutions. A stable protective oxide layer on the metal's surface normally retards the reaction, a circumstance gratefully acknowledged by aeroplane manufacturers.

5.2 Latimer and Volt-equivalent (Frost) Diagrams

There are two main methods of summarizing the thermodynamic stabilities of the oxidation states of elements in aqueous solution, known after their inventors. Latimer and Frost diagrams are usually restricted to the two extremes of standard hydrogen ion (pH = 0) or hydroxide ion (pH = 14) solutions.

5.2.1 Latimer Diagrams

Latimer diagrams were invented by W. M. Latimer and consist of lines of text of the various oxidation states of an element arranged in descending order from left to right, with the appropriate standard reduction potentials (in volts) placed between each pair of states. The diagram for chromium in acid solution is written as:

E versus pH diagrams, invented by Marcel Pourbaix, represent a more comprehensive description of the variations in stabilities of the oxidation states of an element than Latimer or volt-equivalent diagrams. They show areas of stability on plots of E against pH from 0 to 14 for the oxidation states of the element. There is no space in this text to develop Pourbaix diagrams[3] for the elements. Figure 5.1 consists of plots of E against pH; Pourbaix diagrams have Nernst plots for the various couples of an element and include any phase changes, *e.g.* the production of insoluble oxides or hydroxides.

VI		III		II		0
	+1.33		−0.41		−0.91	
$Cr_2O_7{}^{2-}$	——	Cr^{3+}	——	Cr^{2+}	——	Cr

For conciseness in the remainder of the chapter, the Latimer diagrams are presented with the relevant reduction potentials *between* the two oxidation states of the couple, as shown below, or in tabular form.

VI $Cr_2O_7^{2-}$	+1.33	III Cr^{3+}	−0.41	II Cr^{2+}	−0.91	0 Cr

The more exact forms of the aqueous cations, with their primary hydration shells, are normally omitted from the diagrams. As is the case with Cr^{VI}, which in acid solution exists as the dichromate ion, $Cr_2O_7^{2-}$, the forms of any oxo anions are indicated by their formulae in the diagrams. The diagram for chromium summarizes the following important properties:

(i) $Cr_2O_7^{2-}$ is a powerful oxidizing agent, indicated by the high positive value of the E° for its reduction to Cr^{III}.

In practice, chromium does dissolve in hydrochloric and sulfuric acids, but is made passive by concentrated nitric acid. This effect is because of the production of a surface layer of Cr^{III} oxide that is impervious to further acidic attack. It is thought that this is also the reason for chromium (together with nickel) giving stainless steel its non-corroding property.

(ii) Chromium metal should dissolve in dilute [1 mol dm^{-3} H$^+$(aq)] acidic solutions to give the $+3$ state. Written as oxidation reactions (*i.e.* with the signs of the reduction potentials reversed), the positive potentials for the two stages indicate that the reactions would be thermodynamically feasible:

$$Cr(s) + 2H^+(aq) \rightleftharpoons H_2(g) + Cr^{2+}(aq) \quad E^\circ = +0.91 \text{ V}$$

$$Cr^{2+}(aq) + H^+(aq) \rightleftharpoons \tfrac{1}{2}H_2(g) + Cr^{3+}(aq) \quad E^\circ = +0.41 \text{ V}$$

(iii) The Cr^{2+} is unstable in aqueous solution in the presence of dissolved dioxygen. Dioxygen ($O_2/2H_2O$; $E^\circ = +1.23$ V) has the potential to oxidize Cr^{2+} to Cr^{3+}. The $+2$ state is a good reducing agent.

Worked Problem 5.3

Q Construct a fully balanced equation for the oxidation of Cr^{2+} by molecular oxygen, and calculate the standard potential for the process.

A $Cr^{2+}(aq) \rightleftharpoons Cr^{3+}(aq) + e^-$ $E^\circ = +0.41$ V

$O_2(g) + 4H^+(aq) + 4e^- \rightleftharpoons 2H_2O(l)$ $E^\circ = +1.23$ V

$4Cr^{2+}(aq) + O_2(g) + 4H^+(aq) \rightleftharpoons 4Cr^{3+}(aq) + 2H_2O(l)$
$$E^\circ = +1.64 \text{ V}$$

(iv) The $+3$ state, placed between the positive and negative E° values, is the most thermodynamically stable state.

The Latimer diagram for the states of chromium in standard alkaline solution [1 mol dm^{-3} OH$^-$(aq)] is:

VI		III		II		0
	-0.13		-1.1		-1.4	
CrO_4^{2-}	———	$Cr(OH)_3$	———	$Cr(OH)_2$	———	Cr

or in the more concise form:

VI		III		II		0
CrO_4^{2-}	-0.13	$Cr(OH)_3$	-1.1	$Cr(OH)_2$	-1.4	Cr

The information contained by the diagram is:

(i)   Chromium(VI) is the monomeric tetraoxochromate(VI) ion. In alkaline solution, it does not have any oxidant property and is the most stable state of the element under such conditions.
(ii) There is no other solution chemistry, as the $+2$ and $+3$ states exist as solid hydroxides that are easily oxidized to the $+6$ state.

It should be noted that although the thermodynamic data are presented as $E_B{}^\circ$ values, none of the numerical data has been (or could have been) derived by measurements of potentials. The values are derived from thermochemical data and estimates of entropy changes.

Latimer diagrams are available for all the elements except for the unreactive lighter members of Group 18, and form an excellent and concise summary of the aqueous chemistry of such elements. They are used extensively in Chapters 6, 7 and 8. The diagrams used so far in the text consist of one line, but it is sometimes instructive to link oxidation states that are not adjacent to each other and to include the reduction potential appropriate to those states. One example is given in Figure 5.2 for some of the states of iron at pH = 0.

The diagram for the states of iron shows the linkage of the $+3$ and zero states and the $+3/0$ standard reduction potential.

	-0.037	
	$+0.77$	-0.44
Fe^{3+}———	Fe^{2+}———	Fe

Figure 5.2 A Latimer diagram for the +3, +2 and 0 oxidation states of iron; reduction potentials in volts

Worked Problem 5.4

Q   At pH = 0, the following half-reactions contain pairs of oxidation states of the element, M:

$$MO_3^- + 4H^+ + e^- \rightleftharpoons MO^{2+} + 2H_2O \qquad E^\circ = +0.8\,V$$
$$MO^{2+} + 2H^+ + e^- \rightleftharpoons M^{3+} + H_2O \qquad E^\circ = +1.2\,V$$

$$M^{3+} + e^- \rightleftharpoons M^{2+} \qquad E^\circ = +0.2\,V$$
$$M^{2+} + e^- \rightleftharpoons M^+ \qquad E^\circ = -0.8\,V$$
$$M^+ + e^- \rightleftharpoons M \qquad E^\circ = -0.6\,V$$

Convert the information into a Latimer diagram for the element, M, and comment on the redox properties of the various oxidation states.

A The Latimer diagram is:

V	IV	III	II	I	0
	+0.8	+1.2	+0.2	-0.8	-0.6
MO_3^-	MO^{2+}	M^{3+}	M^{2+}	M^+	M

The element, M, should dissolve in dilute acid solution to give the +2 state, which is the most stable state. The reactions:

$$M + H^+ \rightleftharpoons M^+ + \tfrac{1}{2}H_2$$

$$M^+ + H^+ \rightleftharpoons M^{2+} + \tfrac{1}{2}H_2$$

are feasible because their E° values are 0.6 V and 0.8 V, respectively. Any further oxidation reactions by H^+ are not feasible as they have negative E° values.

The +1 state is unstable with respect to **disproportionation** to the +2 state and the neutral element. It has the capacity to oxidize and reduce itself:

$$M^+ \rightleftharpoons M^{2+} + e^- \qquad E^\circ = +0.8 \text{ V}$$

$$M^+ + e^- \rightleftharpoons M \qquad E^\circ = -0.6 \text{ V}$$

$$2M^+ \rightleftharpoons M^{2+} + M \qquad E^\circ = +0.8 - 0.6 = 0.2 \text{ V}$$

The positive E° value indicates that the disproportionation reaction is feasible.

By similar reasoning, the +4 state is unstable with respect to disproportionation to the +5 and +3 states:

$$MO^{2+} + 2H_2O \rightleftharpoons MO_3^- + 4H^+ \qquad E^\circ = -0.8 \text{ V}$$

$$MO^{2+} + 2H^+ \rightleftharpoons M^{3+} + H_2O \qquad E^\circ = +1.2 \text{ V}$$

$$2MO^{2+} + H_2O \rightleftharpoons MO_3^- + M^{3+} + 2H^+ \quad E^\circ = +1.2 - 0.8 = 0.4 \text{ V}$$

To oxidize the element to beyond its +2 state requires an oxidizing agent of suitable potential, *i.e.* one with an E° value greater than +0.2 V for the +2 to +3 oxidation, and one with an E° value greater than +1.2 V for the +3 to +4 oxidation.

5.2.2 Volt-equivalent Diagrams (Frost Diagrams)

Equation (4.5) relates the standard reduction potential (volts) to the standard Gibbs energy (in J mol^{-1}). This equation can be rearranged:

$$\Delta G^{\oplus}/F = -nE^{\oplus} \qquad (5.11)$$

The value of the standard reduction potential is multiplied by n (the number of electrons participating in the reduction and the product represents, with its sign changed, the **volt-equivalent** of the **change in standard Gibbs energy** for the reduction process. Equation (5.11) may be expanded to include the Gibbs energies of the oxidized and reduced states for a general reduction process:

$$\Delta G^{\oplus}/F = [G^{\oplus}(\text{reduced state}) - G^{\oplus}(\text{oxidized state})]/F$$
$$= G^{\oplus}(\text{reduced state})/F - G^{\oplus}(\text{oxidized state})/F = -nE^{\oplus} \quad (5.12)$$

Figure 5.3 shows a plot of $G^{\oplus}/F$ against oxidation state for the reduction process, M^{3+}/M^{+}, for a couple with a positive $E^{\oplus}$ value ($+1.1$ V).

The $E^{\oplus}$ value is asserted to be positive, so the accompanying value of $\Delta G^{\oplus}$ is negative; the value of $G^{\oplus}/F$ for the oxidized state of the couple is higher by 2×1.1 V than that for the reduced state. A volt-equivalent consists of a plot of $G^{\oplus}/F$ values against the oxidation state for ions of the element under consideration. Ions of less stability are placed higher up the $G^{\oplus}/F$ axis in the volt-equivalent diagram; those with greater stability are placed lower down.

The plot of the volt-equivalent ($G^{\oplus}/F$) against oxidation state for chromium in acid solution is shown in Figure 5.4, with the oxidation state decreasing from left to right. The individual steps from left to right are reduction processes. In this way, Frost diagrams can be associated easily with the corresponding Latimer diagram. The origin of such a graph is the volt-equivalent of *zero* for the zero oxidation state of the element, consistent with the Gibbs energy of formation of an element in its standard state having the conventional zero value. The point for Cr^{2+} is placed at oxidation state 2, 1.82 V lower down than the metal. This is because the volt-equivalent value of the reduction potential of -0.91 V is given by $-\Delta G^{\oplus}/F = (2 \times -0.91 \text{ V}) = -1.82$ V or $\Delta G^{\oplus}/F = 1.82$ V, indicating that the neutral element is more stable than Cr^{2+}.

The point for Cr^{3+} is placed a further 0.41 V down from -1.82 V at -2.23 V, indicating the extra stability of the Cr^{III} state. The point for $Cr_2O_7^{2-}$ is placed $1.33 \times 3 = 3.99$ V higher than that for the $+3$ state, indicating the instability of the former state with respect to the latter. The factor of three is used because the $+6$ to $+3$ conversion is a three-electron change ($n = 3$). Caution is required with this placement because the form of the Cr^{VI} ion contains two chromium atoms, $Cr_2O_7^{2-}$. This is because reduction potentials refer to the Gibbs energy per electron, so it is important when potentials are converted to Gibbs energy changes that attention is given to the number of moles of the element participating in every half-reaction.

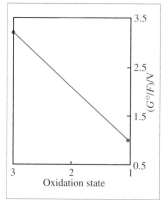

Figure 5.3 A plot of $G^{\oplus}/F$ against oxidation state for the reduction process M^{3+}/M^{+}

Volt-equivalent diagrams were proposed and first used by Arthur A. Frost[1] and modified for general use by C. S. G. Phillips and R. J. P. Williams.[2]

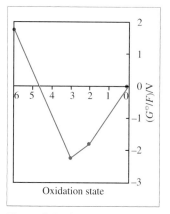

Figure 5.4 A volt-equivalent diagram for chromium at pH = 0

Volt-equivalent diagrams convey the same amount of information as do Latimer diagrams about the relative stabilities of the oxidation states of an element and their oxidation/reduction properties, but do it in a graphical manner. Such diagrams are given in subsequent chapters for selected elements to illustrate further the differences in potentials between successive oxidation states.

The theory of reduction potentials and their understanding in terms of the stages of reduction processes, as described above, are used in the discussion of the periodicity of the ions formed by the elements and their redox interactions in the subsequent chapters.

Summary of Key Points

1. The Nernst equation was introduced and its relevance to the dependence of reduction potentials upon the pH of the solution was described.

2. The theoretical and practical limits of reduction potentials were defined. Outside such limits the oxidized form or the reduced form of a couple is unstable with respect to reaction with water.

3. The basis and interpretation of Latimer diagrams were dealt with.

4. The construction and interpretation of volt-equivalent diagrams were described.

References

1. A. A. Frost, *J. Am. Chem. Soc.*, 1951, **73**, 2680.
2. C. S. G. Phillips and R. J. P. Williams, *Inorganic Chemistry*, vol. 1, Oxford University Press, Oxford, 1965, p. 314.
3. J. A. Campbell and R. A. Whiteker, *A Periodic Table Based on Potential–pH Diagrams*, in *J. Chem. Ed.*, 1969, **46**, 92.

Further Reading

W. M. Latimer, *Oxidation Potentials*, Prentice-Hall, New York, 1952.
D. R. Rosseinsky, *Electrode Potentials and Hydration Energies, Theories and Correlations*, in *Chem. Rev.*, 1965, **65**, 467.
M. Pourbaix, *Atlas of Electrochemical Equilibria in Aqueous Solutions*, National Association of Corrosion Engineers, Houston, 1974.

Problems

5.1. Write fully balanced equations for the half-reaction reductions of the couples: (i) $H_3AsO_4/HAsO_2$; (ii) $HBrO/Br^-$; (iii) $HOClO/Cl^-$ and (iv) MnO_4^-/MnO_2.

5.2. The standard reduction potentials for the half-reactions in Problem 5.1 are respectively $+0.56$, $+1.33$, $+1.57$ and $+1.68$ V. Derive Nernst equations for the four half-reactions, paying particular attention to the "Nernst slopes".

5.3. From the Latimer diagram for the element M, calculate the standard reduction potential for the M^{2+}/M reduction, and comment on the stability of the $+1$ state in aqueous solution:

$$
\begin{array}{ccccc}
\text{II} & & \text{I} & & 0 \\
M^{2+} & -0.2\ \text{V} & M^+ & +0.6\ \text{V} & M
\end{array}
$$

5.4. From the values of the standard reduction potentials of the following couples, comment on the stability of the oxidized forms and the reduced forms of the elements:

(i) O_3/O_2 $E^\circ = +2.08$ V at pH $= 0$
(ii) SO_4^{2-}/SO_3^{2-} $E^\circ = -0.93$ V at pH $= 14$

5.5. The Cr^{2+} ion can be produced in acid solution by the reduction of Cr^{3+} by metallic zinc. Check that this is so from the information in the text, and calculate a value for the reaction potential. Is the Cr^{2+} stable in acidic solution?

5.6. The reduction of Cr^{VI} to Cr^{III} in 1 mol dm^{-3} H^+ solution has a potential of 1.33 V. Write a fully balanced half-reaction for the reduction process and calculate the value of the potential in 0.1 mol dm^{-3} H^+. Assume that no change occurs to the form of the Cr^{VI} ion, which exists as $Cr_2O_7^{2-}$.

6
Periodicity of Aqueous Chemistry I: s- and p-Block Chemistry

The subject of this chapter is the periodicity of the aqueous chemistry of the elements of the s-block (Groups 1 and 2) and the p-block (Groups 11–18) of the Periodic Table. Modified Latimer diagrams summarize the chemistry of all the elements, and some volt-equivalent diagrams are given to represent the inter-relations between various oxidation states of the elements. Explanations of some trends in redox chemistry are discussed in detail.

Aims

By the end of this chapter you should understand:

- The periodicity in the ions formed by the s- and p-block elements
- The variations in the values of standard reduction potentials of the Group 1 M^+/M couples (M = Li, Na, K, Rb, and Cs)
- The trend shown by the standard reduction potentials for the reductions of sodium, magnesium and aluminium ions to their respective metals
- The general discussions of the aqueous chemistry of the elements of Groups 13–17 and that of xenon from Group 18

6.1   Periodicity of the Ions Formed by the s- and p-Block Elements

6.1.1   Valency and Oxidation State: Differences of Terminology

The **valency** of an element is its **combining power,** represented as the number of electron-pair bonds in which it can participate by sharing some or all of its valency electrons with an *equal number* from the atoms to which it is bonded. This number is equal to or sometimes less than the number of **valency electrons,** *i.e.* those electrons that are capable of entering into covalent bonding. These are the electrons in the **valence shell,** *i.e.* those in the "outer" electronic configuration of the atom. For example, the sulfur atom has an outer electronic configuration $3s^2 3p^4$, and could be expected to be six-valent if all the valence electrons took part in covalent bonding, as in SF_6. In such a case the central atom is **covalently saturated.** Sulfur exerts valencies that are less than the maximum, as in the compounds SF_4 (four-valent S), SCl_2 (two-valent S) and H_2S (two-valent S).

Ionic compounds have indications of the "valency" of the constituent elements in their stoichiometries, *e.g.* in NaCl both atoms are monovalent, but in MgF_2 the magnesium is divalent in combining with the monovalent fluorine. In such compounds the term "valency" has to be used with caution. In NaCl, each ion has a coordination number of six, *i.e.* each ion has six nearest neighbours that might be regarded as being ionically bonded to the central ion. In the structure of MgF_2 (which has the rutile structure of TiO_2), each magnesium ion is surrounded by six fluoride ions, and each fluoride ion is surrounded by three magnesium ions arranged in a trigonal plane. The terms monovalent and divalent, for Na and Mg respectively, are used to describe the metals in the compounds mentioned, but the **oxidation state** concept is more accurate and understandable.

There are important differences between the two concepts that should be appreciated. Oxidation state is an imaginary charge on an atom that is in combination with one of the very electronegative atoms, *i.e.* F or O, in which their oxidation states are deemed to be -1 and -2 respectively. For example, in MgF_2 and MgO the oxidation state of the magnesium is $+2$ (*i.e.* II in the *Roman numerals* conventionally used to indicate oxidation states, as Mg^{II}) and equal to the charge on the Mg^{2+} ions in those compounds. In ionic compounds there are no discrete electron-pair bonds so the strict definition of valency, given above, does not apply.

In essentially covalent compounds the oxidation state concept can be useful in their classification. The two main oxides of sulfur are SO_2 and SO_3 and, if the oxidation state of oxygen is taken to be -2 by convention, the oxidation states of the sulfur atoms in the oxides are $+4$ and $+6$ respectively, *i.e.* they contain S^{IV} (sulfur-4) and S^{VI} (sulfur-6).

Worked Problem 6.1

Q   Explain why the oxidation state of oxygen in the oxide ion is taken to be -2.

A   In fully ionic oxides the oxygen species present is O^{2-}; therefore the oxidation state of the oxygen is equal to the charge, *i.e.* -2. This value is taken to be the oxidation state of oxygen in compounds that are not necessarily ionic. The oxide ion has the fully occupied valence orbitals equivalent to the electronic configuration of neon, $2s^2 2p^6$, the Group 18 element at the end of the same period.

Worked Problem 6.2

Q   Assign oxidation states to the elements in the compounds H_2O_2 and FeS_2 (iron pyrites).

A   These assignments depend on some chemical knowledge of the structure of the compound. The atom–atom linkages in hydrogen peroxide are $H-O-O-H$. If the hydrogen atoms are considered to be in their $+1$ states, that leaves -2 to be shared between the oxygen atoms, making them each -1. If the sulfur atoms were to be given oxidation states of -2 as in normal sulfides, *e.g.* Na_2S, the oxidation state of the iron would be $+4$. Normally iron exists in its $+2$ and $+3$ states in Nature, and assuming that the iron is in the $+2$ state makes the S_2^{2-} have a -2 state, *i.e.* both sulfur atoms are in state -1. There is an $S-S$ linkage in the disulfide ion, similar to the $O-O$ linkage in hydrogen peroxide.

The oxidation states of sulfur in the compounds SF_4 and SCl_2, $+4$ and $+2$ respectively, may be included in their formulae as right-hand superscripts in Roman numerals: $S^{IV}F_4$ and $S^{II}Cl_2$. This convention is used whenever the oxidation state of an element needs to be emphasized. For example, the compound Pb_3O_4 may be regarded as a mixed oxide, $PbO_2.2PbO$, and could be written as $Pb^{IV}Pb_2^{II}O_4$. Negative oxidation states are rarely indicated in formulae.

The sulfur atoms in the compounds SF_4, SCl_2 and H_2S have valencies of 4, 2 and 2, but, if the formal oxidation states of the fluorine, chlorine and hydrogen atoms are taken to be -1, -1 and $+1$, respectively, the oxidation states of the sulfur atoms in those compounds are $+4$, $+2$ and -2.

In H_2S the hydrogen atom is assigned the oxidation state of $+1$ because hydrogen is less electronegative than sulfur. In a compound such as LiH (which is ionic, Li^+H^-) the hydrogen is in oxidation state -1, consistent with H being more electronegative than Li. In the chlorate(VII) ion, ClO_4^-, the chlorine atom has an oxidation state of $+7$ because the four oxygen atoms are considered to be in the -2 state, and a central chlorine(VII) makes the overall charge minus one. The bonding in the ion is essentially covalent, and the massive amount of energy which would have to be used to remove (*i.e.* ionize) the seven valency electrons of the chlorine atom is sufficient to prevent any ionic bonding between the chlorine atom and the four oxygen atoms.

6.1.2 The s- and p-Block Elements other than Hydrogen

Table 6.1 gives the variations of the positive oxidation states of the elements of the s- and p-blocks of the Periodic Table. The elements of the 2nd period, Li–Ne, show the values expected from the strict application of the octet rule.

The octet rule indicates that the oxidation state of an element in a compound is that achieved when the element loses electrons to become an ion with an electronic configuration identical to the Group 18 element in the previous period $[(n-1)s^2(n-1)p^6]$, or gains electrons to become an ion with an electronic configuration identical to that of the Group 18 element in the same period (ns^2np^6). The necessary changes are summarized in Table 6.2.

Table 6.1 The positive oxidation states of the elements of the s- and p-blocks that are stable in aqueous solution

Group 1	2	13	14	15	16	17	18
Li	Be	B	C	N	O	F	Ne
				+5			
			+4				
		+3		+3			
	+2						
+1							
Na	Mg	Al	Si	P	S	Cl	Ar
						+7	
					+6		
				+5		+5	
			+4		+4		
		+3		+3		+3	
	+2				+2		
+1						+1	
K	Ca	Ga	Ge	As	Se	Br	Kr
						+7	
					+6		
				+5		+5	
			+4		+4		
		+3		+3		+3	
	+2				+2		+2
+1		+1				+1	
Rb	Sr	In	Sn	Sb	Te	I	Xe
							+8
						+7	
					+6		+6
				+5		+5	
			+4		+4		+4
		+3		+3		+3	
	+2		+2		+2		+2
+1		+1				+1	

(continued)

Table 6.2 Electronic changes compatible with the octet rule

Group	Outer electronic config.	Electron change required to produce a Group 18 config.
1	s^1	-1
2	s^2	-2
13	s^2p^1	-3
14	s^2p^2	±4
15	s^2p^3	$+3$
16	s^2p^4	$+2$
17	s^2p^5	$+1$
18	s^2p^6	0

Table 6.1 *continued*

Group 1	2	13	14	15	16	17	18
Cs	Ba	Tl	Pb	Bi	Po	At	Rn
							+6?
				+5			
			+4		+4		
		+3		+3			
	+2		+2		+2		+2
+1		+1					

The most spectacular variations from the octet rule are in xenon chemistry, where the rule would predict zero valency. Xenon forms a small range of compounds with oxygen and fluorine in which it expands its valence shell to produce oxidation states of +2, +4, +6 and +8.

The elements of the subsequent periods follow the rule, but there are many exceptions. Some elements form compounds that involve **hypervalency** (dealt with in Section 6.2.6) in which the valency of the central atom is higher than that expected from the octet rule, *e.g.* ClO_4^- has chlorine in its $+7$ state, which formally forms one single bond to O^- and participates in three double Cl=O bonds, making the formal valency equal to seven. Other compounds contain ions that have oxidation states two units lower than expected from the octet rule, *e.g.* Sn^{2+}, and which exhibit the **inert pair effect** (dealt with in Section 6.2.3).

In Groups 15–17 there are examples of compounds and ions in which the oxidation states of the group elements are negative. These are given in Table 6.3.

Table 6.3 Negative oxidation states of Group 15–17 elements in compounds and ions

Group 15	Oxidation state	16	Oxidation state	17	Oxidation state
NH_2OH	-1	H_2O_2	-1	HF	-1
NH_3OH^+	-1	HO_2^-	-1	F^-	-1
N_2H_4	-2	H_2O	-2	Cl^-	-1
$N_2H_5^+$	-2	HO^-	-2	Br^-	-1
NH_3	-3	H_2S	-2	I^-	-1
NH_4^+	-3	HS^-	-2		
PH_3	-3	S^{2-}	-2		
P_2H_4	-2	H_2Se	-2		
		HSe^-	-2		
		Se^{2-}	-2		
		H_2Te	-2		
		HTe^-	-2		
		Te^{2-}	-2		

Across the elements of Groups 1, 2 and 13 to 18 there are major changes in the types of ions that are stable in solution and in their redox behaviour. There is a change from the electropositive elements of the first two groups, which are powerful reducing agents easily oxidized to their $+1$ and $+2$ states, respectively, to the elements of Group 17 which are oxidizing agents, ranging from the powerful oxidizers, F_2, Cl_2 and Br_2, to I_2, which is not as powerful. Down any group, the reducing powers of the electropositive elements increase in general and the oxidizing power of the electronegative elements decrease. The details of these generalizations are presented in tables that are modified Latimer-type diagrams and in some volt-equivalent diagrams.

The forms of the aqueous ions of the s- and p-blocks of the Periodic Table vary from the positively charged $+1$, $+2$ and $+3$ ions typical of the elements of Groups 1, 2 and 13, respectively, e.g. Na^+, Mg^{2+} and Al^{3+}, to the oxoanions typical of the positive oxidation states of the elements of Groups 14 to 18, e.g. CO_3^{2-}, NO_3^-, PO_4^{3-}, SO_4^{2-} and ClO_4^-, some of which have central atoms that are hypervalent. The stability in aqueous solution of positive ions depends largely on the balance between the ionization energy expended to attain the positive charge and the compensating enthalpy of hydration of the ion. Providing the hydration enthalpy outweighs the ionization energy, the positive ion is probably stable in aqueous solution. If the two quantities are reversed in magnitude, the existence of positive ions is not possible and the production of hydrated oxoanions occurs. Such an argument ignores the effects of entropy changes, but these are relatively minor.

The oxoanions may be thought of as resulting from the hydrolysis of the positively charged ions, with the formation of double covalent bonds to neutral oxygen atoms and single covalent bonds to oxygen atoms possessing a formal single negative charge. The production of single and double bonds to oxygen atoms coupled with the enthalpy of hydration of the oxoanion confer stability on the ion in aqueous solution. As the oxidation state of the central atom increases, the electronegativity of the atom increases and this makes covalency more likely. In acidic solutions, oxoanions in general become less stable as the oxidation state of the central ion increases and their oxidizing power increases. Hydrolysis is encouraged in alkaline solution, and this stabilizes the oxoanions with higher oxidation states of the central ion, which then tend to be less powerful oxidants or lose their oxidative capacity altogether.

Note that the elements that show a variable valency or oxidation state do so by two-unit changes. This occurs when two electrons are unpaired so that both may participate in bonding in the higher state. This is a major difference between the s- and p-block elements and those of the d-block, which show one-unit changes in their oxidation states.

6.2 The Detailed Aqueous Chemistry of the s- and p-Block Elements

In this section the aqueous chemistry of the s- and p-block elements, their ions and redox chemistry are described and discussed.

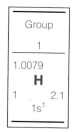

For simplicity, the hydrated proton is written as $H^+(aq)$ in subsequent reactions and half-reactions unless there is a particular reason to formulate it as $H_3O^+(aq)$. The value of ΔH° for the half-reaction (6.1) is derived from those of:

$$-\Delta_{hyd}H^\circ 1(H^+, g) - I_1(H) - 6.2$$
$$-\Delta_a H^\circ = -426.2 \text{ kJ mol}^{-1}$$

Element "boxes" like the one above are shown alongside their descriptive chemistry sections in this and the next chapters. The element symbols are shown with the relative atomic mass as a left superscript, the atomic number as a left subscript, and the Allred–Rochow electronegativity coefficient as a right subscript. The valence shell electronic configuration is shown below the element symbol.

6.2.1 Hydrogen

The first element, hydrogen, has an Allred–Rochow electronegativity coefficient of 2.1, and an electronic configuration $1s^1$. The atom may lose the single electron to become a proton, which exists in aqueous solutions as the hydroxonium ion, $H_3O^+(aq)$, in which the proton is covalently bonded to the oxygen atom of a water molecule. The ion is hydrated, as is discussed extensively in Chapter 2. The reduction of the hydrated proton by an electron forms the reference standard half-reaction for the scale of reduction potentials:

$$H^+(aq) + e^- \rightleftharpoons \tfrac{1}{2}H_2(g) \qquad (6.1)$$

The values of ΔG° and ΔH° for the half-reaction are taken to be zero by convention, but the value of ΔH° in absolute terms is $-426.2 \text{ kJ mol}^{-1}$, as shown in Section 6.2.2, and that of ΔG° is $-445.7 \text{ kJ mol}^{-1}$, as may be derived from the data given in Figure 4.2.

Worked Problem 6.3

Q Calculate the value of $T\Delta S^\circ$ and ΔS° for half-reaction (6.1) at 298.15 K from the data given in the text.

A $T\Delta S^\circ = (\Delta H^\circ - \Delta G^\circ) = -426.2 + 445.7 = -19.5 \text{ kJ mol}^{-1}$
$\Delta S^\circ = (\Delta H^\circ - \Delta G^\circ)/298.15 = (-426.2 + 445.7)/298.15$
$\quad = 65.4 \text{ J K}^{-1}\text{mol}^{-1}$

The hydrated proton is the simplest (in that it is monatomic) and strongest acid in the aqueous system. Any compound that is a stronger acid decomposes in aqueous solution to give a solution containing hydrated protons and counter-anions, e.g. HCl dissociates to give hydrated protons and hydrated chloride ions.

The Group 1 elements, and the Group 2 elements except for the lighter members, Be and Mg, form ionic compounds with hydrogen in which the hydrogen occurs as the hydride ion, H^-. In these cases the hydrogen atom has an attached electron that gives the hydride ion, which has the helium electronic configuration, $1s^2$. The compounds react with water because the hydride ion is unstable in its presence:

$$H^-(aq) + H_2O(l) \rightarrow H_2(g) + OH^-(aq) \qquad (6.2)$$

Thus, sodium hydride, NaH, reacts with water to give a solution of sodium hydroxide. In terms of acid/base theory, the hydride ion is a stronger base than OH^- and readily accepts a proton from water.

The Latimer diagram for hydrogen is:

$$
\begin{array}{ccc}
+1 & 0 & -1 \\
\end{array}
$$

$$
H^+ \quad 0 \quad {}^{1\!/}_{2}H_2 \quad -2.23 \quad H^-
$$

6.2.2 Groups 1 and 2 Metals

Apart from hydrogen, which is treated separately, the elements of Groups 1 and 2 are all metallic in character and form hydrated $M^+(aq)$ and $M^{2+}(aq)$ hydrated ions, respectively, in aqueous solution. The ions formed are those expected from the octet rule, the elements of both Groups losing their valency electrons to give the electronic configuration of the Group 18 element in the previous period. Table 6.4 gives the standard reduction potentials for the metallic elements of Groups 1 and 2 at pH = 0.

Dilute aqueous solutions of strong acids (*e.g.* HCl or H_2SO_4) contain sufficient concentrations of hydrated protons to oxidize many metals, to produce their most stable states in solution. The only thermodynamic condition for metal oxidation is that the reduction potential of the metal ion produced should be *negative*. In general, for the metal ion M^{n+} undergoing reduction to the metal, if the standard reduction potential for the half-reaction:

$$
M^{n+}(aq) + ne^- \rightleftharpoons M(s) \tag{6.3}
$$

is negative, *e.g.* $-x$ V, the potential for the reverse half-reaction:

$$
M(s) \rightleftharpoons M^{n+}(aq) + ne^- \tag{6.4}
$$

is positive, $+x$ V. The reference half-reaction has a potential of zero:

$$
nH^+(aq) + ne^- \rightleftharpoons {}^{n\!/}_{2}H_2(g) \tag{6.5}
$$

The overall reaction, constructed by adding together the two half-reactions so that the electrons cancel out, is:

$$
M(s) + nH^+(aq) \rightleftharpoons M^{n+}(aq) + {}^{n\!/}_{2}H_2(g) \tag{6.6}
$$

This overall reaction has a positive potential, $+x$ V, a negative change in Gibbs energy, and therefore is feasible. Whether such reactions occur depends upon the absence of an oxide layer that might prevent the oxidant reaching the metal surface.

The very negative reduction potentials of the Groups 1 and 2 elements indicate that the reductions of their ions to the elements is very Gibbs-energy consuming, and that the reverse processes, where the elements provide electrons for reducing purposes, are extremely favoured thermodynamically.

Group 1 metals	Group 2
6.941 **Li** 3 1.0 $2s^1$	9.012 **Be** 4 1.5 $2s^2$
22.990 **Na** 11 1.0 $3s^1$	24.305 **Mg** 12 1.2 $3s^2$
39.098 **K** 19 0.9 $4s^1$	40.078 **Ca** 20 1.0 $4s^2$
85.468 **Rb** 37 0.9 $5s^1$	87.62 **Sr** 38 1.0 $5s^2$
132.905 **Cs** 55 0.9 $6s^1$	137.327 **Ba** 56 1.0 $6s^2$
$(223.02)^a$ **Fr** 87 0.9 $7s^1$	226.025 **Ra** 88 1.0 $7s^2$

[a]Most stable isotope

Table 6.4 The standard reduction potentials for the metallic elements of Groups 1 and 2 at pH = 0

Couple	$E°/$ V
Li^+/Li	-3.04
Na^+/Na	-2.71
K^+/K	-2.92
Rb^+/Rb	-2.99
Cs^+/Cs	-3.03
Be^{2+}/Be	-1.85
Mg^{2+}/Mg	-2.37
Ca^{2+}/Ca	-2.87
Sr^{2+}/Sr	-2.89
Ba^{2+}/Ba	-2.91
Ra^{2+}/Ra	-2.92

Thermodynamic feasibility of any reaction is not necessarily paralleled by the rate of the reaction. The rate of a reaction, however feasible the reaction is thermodynamically, is governed by the magnitude of the energy barrier to reaction represented by the **activation energy**, *i.e.* the energy required to elevate the reacting mixture to its **transition state** from which it may proceed to give products or return to its initial state.

All the Groups 1 and 2 metals have reduction potentials with respect to their respective M^+(aq) and M^{2+}(aq) ions that are extremely negative. They fall into the region below the lower practical boundary of Figure 5.1, so they are expected to react with neutral liquid water. In its reactions with liquid water, lithium is relatively slow compared to the other Group 1 metals, which react very rapidly (Cs almost explosively) with water. The Group 2 elements react at very different rates with water. Beryllium only reacts at red heat. Magnesium reacts slowly with cold water to give dihydrogen and the insoluble hydroxide, $Mg(OH)_2$. The other members of the group react quite vigorously with water, but not as rapidly as the heavier Group 1 metals.

The high reducing capacity of the Group 1 elements is detailed above. Those of the heavier Group 2 elements (Ca, Sr, Ba and Ra) are almost as high as the Group 1 elements. Although Be and Mg are still powerful reducing agents, their less negative potentials arise because of the greater energies required to cause their double ionization. This point is dealt with in detail later in this section.

The E° Values of the Group 1 M^+/M Reductions

The accepted values for the standard reduction potentials for the Group 1 unipositive cations being reduced to the solid metal are given in Table 6.4, and some thermochemical data and the appropriate ionic radii are given in Table 6.5.

Table 6.5 Data for Group 1 elements, M, and their cations, M^+; the units for all the energies are kJ mol^{-1}

	Li	Na	K	Rb	Cs
Δ_aH° (M, g)	+159	+108	+89	+81	+77
I_1(M)	+520	+496	+419	+403	+376
$\Delta_{hyd}H^\circ$(M^+, g)	−538	−424	−340	−315	−291
r_{ionic}(M^+)/pm	76	102	138	152	167

The E° values imply that all the elements are very powerful reducing agents (M^+ being difficult to reduce to M) and that lithium and caesium are the most powerful, with sodium being relatively the least powerful. The reducing powers of the elements K and Rb are intermediate between those of Li and Cs and that of Na. The following calculations identify the reasons for the very negative values of E°.

The calculations that follow are carried out in terms of *enthalpy changes*, with entropy changes being ignored. This is justified because the entropy changes are (a) expected to be fairly similar for the reactions of the elements of the same periodic group, (b) relatively small compared to the enthalpy changes, and (c) somewhat more difficult to estimate.

The standard enthalpy change for the reduction of $M^+(aq)$ to $M(s)$ may be estimated from the above data by carrying out a calculation using the thermochemical cycle shown in Figure 6.1 for the overall reaction:

$$M^+(aq) + \tfrac{1}{2}H_2(g) \rightleftharpoons H^+(aq) + M(s) \tag{6.7}$$

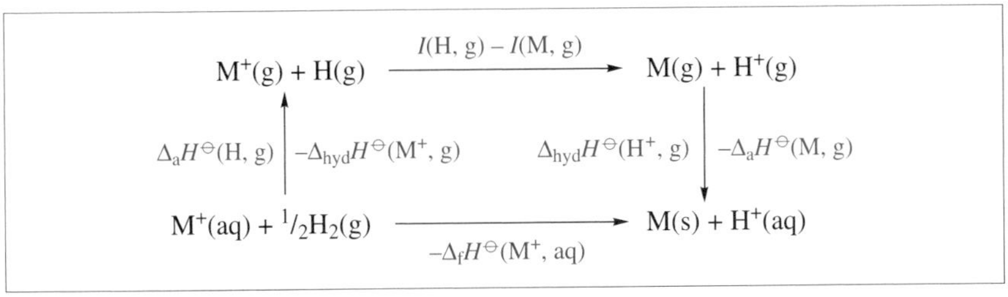

Figure 6.1 A thermochemical cycle for the reduction of aqueous M^+ ions to the metal M by dihydrogen

The enthalpy change of reaction (6.7) is given by the equation:

$$\Delta H^{\circ}(6.7) = -\Delta_f H^{\circ}(M^+, \text{ aq}) = -\Delta_{\text{hyd}} H^{\circ}(M^+, \text{ g}) - I_1(M)$$

$$- \Delta_a H^{\circ}(M, \text{ g}) + \Delta_a H^{\circ}(H, \text{ g}) + I_1(H) + \Delta_{\text{hyd}} H^{\circ}(H^+, \text{ g})$$

$$= -\Delta_{\text{hyd}} H^{\circ}(M^+, g) - I_1(M) - \Delta_a H^{\circ}(M, \text{ g}) + 420 \tag{6.8}$$

The reduction potentials of the Group 1 elements are calculated from equation (6.8) by dividing the enthalpy change by $-F$. The results are given in Table 6.6.

The last three terms of equation (6.8) amount to 420 kJ mol^{-1}, as shown in Section 6.2.1.

The calculated values for the reduction potentials are very similar to those observed, the small differences being attributable to the ignored entropy terms. The calculation of reduction enthalpy shows that the value is the resultant (a relatively small quantity) of the interaction between two numerically large quantities: the ionization energy of the gaseous metal atom and the hydration energy of the cation, with the enthalpy of atomization making a relatively small contribution. The trends in the values of the three quantities are understandable qualitatively in terms of: (i) the changes in size of the ions (the enthalpy of hydration decreases with increasing ionic radius, see the data in Table 6.5); (ii) the electronic configuration of the atoms (ionization energy decreases as Z increases); and (iii) the strength of the metallic bonding, which becomes weaker as atomic size increases (the effective nuclear charge decreases). The values of the enthalpies of hydration of the cations, the ionization energies of the atoms, and the enthalpies of atomization of the Group 1 elements are plotted against the ionic radii of the elements in Figure 6.2.

Table 6.6 Calculated reduction enthalpies and E° values for the Group 1 cations

Cation/ element	Reduction enthalpy/ kJ mol^{-1}	E°/V
Li$^+$/Li	279	− 2.89
Na$^+$/Na	240	− 2.49
K$^+$/K	252	− 2.61
Rb$^+$/Rb	251	− 2.60
Cs$^+$/Cs	258	− 2.67

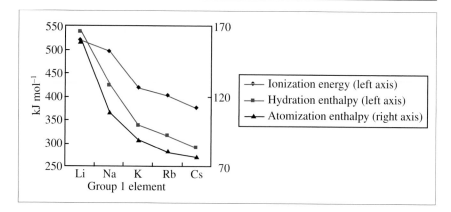

Figure 6.2   Plots of atomization enthalpy, ionization energy and the negative value of the enthalpy of hydration for the Group 1 elements and ions

It can be seen from Figure 6.2 that the widening gap between the ionization energy and negative values of the enthalpy of hydration, which would make the Group 1 elements have greater reducing power down the group, is offset by the trend in atomization energy. Thus, the latter trend in a relatively small quantity serves to maintain the reducing powers of the Group 1 elements at about the same level.

Worked Problem 6.4

Q   Calculate the change in entropy for reaction (6.7) with M = Li, and calculate the value of $T\Delta S^\circ$ Compare the value of $T\Delta S^\circ$ with that of ΔH° for the process and comment on the relative importance of the two values in determining E° for the Li^+/Li reduction.

A   The cycle of Figure 6.1 suggests that the equation giving the entropy change for reaction (6.7) is, when combined with the result of Worked Problem 6.3:

$$
\begin{aligned}
\Delta S^\circ(6.7) &= -\Delta_{hyd}S^\circ(M^+,\, g) + S^\circ(M,\, g) - S^\circ(M^+,\, g) \\
&\quad - \Delta_a S^\circ(M,\, g) - 65.4 \\
&= -\Delta_{hyd}S^\circ(M^+,\, g) + S^\circ(M,\, g) - S^\circ(M^+,\, g) \\
&\quad + S^\circ(M,\, s) - S^\circ(M,\, g) - 65.4 \\
&= -\Delta_{hyd}S^\circ(M^+,\, g) - S^\circ(M^+,\, g) + S^\circ(M,\, s) - 65.4 \\
&= 140.6 - 133.1 + 29.1 - 65.4 = -29.0 \text{ J K}^{-1}\,\text{mol}^{-1}
\end{aligned}
$$

The value of $T\Delta S^\circ(6.7) = -(29 \times 298.15)/1000 = -8.6 \text{ kJ mol}^{-1}$
   Table 6.6 gives $\Delta H^\circ(6.7) = 279 \text{ kJ mol}^{-1}$, so the $T\Delta S^\circ$ term represents only 3% of the contribution to the E° value. Inclusion of the $T\Delta S^\circ$ term gives a Gibbs energy change for the Li^+/Li reduction

of $279 - (-8.6) = 287.6 \text{ kJ mol}^{-1}$ that converts to a calculated E° value of $-287600/96485 = -2.98$ V, not very far removed from the experimental value of -3.04 V.

6.2.3 The Aqueous Chemistry of the Group 13 Elements

Table 6.7 contains data for the Group 13 elements at pH values of 0 and 14, arranged in the form of Latimer-type diagrams for each element at the two pH values. The appropriate standard potentials appear between the two relevant oxidation states of the element.

Table 6.7   Standard reduction potentials for the Group 13 elements at pH = 0 and pH = 14

E°/V at pH = 0				E°/V at pH = 14		
III		I		0	III	0
$B(OH)_3$			-0.87	B	$B(OH)_4^-$	-1.81	B
Al^{3+}			-1.66	Al	$Al(OH)_4^-$	-2.33	Al
Ga^{3+}			-0.55	Ga	$GaO(OH)_2^-$	-1.22	Ga
In^{3+}	-0.44	In^+	-0.14	In		
Tl^{3+}	$+1.25$	Tl^+	-0.34	Tl		

The data in Table 6.7 show the transition down the group from the non-metallic behaviour of boron, through the amphoteric Al and Ga metals, to the metallic In and Tl.

In acidic solution In and Tl have $+1$ states, consistent with the **inert pair effect** that affects the heavier elements of Groups 13 to 15.

Group 13
10.811
B
5   2.0
$2s^2 2p^1$
26.982
Al
13   1.5
$3s^2 3p^1$
69.723
Ga
31   1.8
$3d^{10}4s^2 4p^1$
114.818
In
49   1.5
$4d^{10}5s^2 5p^1$
204.383
Tl
81   1.4
$4f^{14} 5d^{10}6s^2 6p^1$

Box 6.1   Relativity Theory and the Inert Pair Effect

The values of ionization energies and atomic sizes are influenced by **relativistic effects** that, for valence electrons, increase with the value of Z^2, and become sufficiently important in the elements of the 6th period (Cs–Rn) to explain largely their chemical differences from the elements of the 5th period (Rb–Xe). The initial relativistic effect is to cause a decrease in the radius of the 1s atomic orbital of the atom. The mass of the electron in the 1s orbital becomes higher as the nuclear charge increases because the velocity of the electron increases.

Since the radius of the 1s orbital is inversely proportional to the mass of the electron, the radius of the orbital is reduced compared to that of the non-relativistic radius. This s-orbital contraction affects

According to relativity theory, the mass of a particle, moving with a velocity v, is given by:

$$m = \frac{m_0}{\sqrt{1 - \frac{v^2}{c^2}}} \qquad (6.9)$$

where m_0 is the rest mass of the particle, and c is the velocity of electromagnetic radiation.

the radii of all the other orbitals in the atom up to, and including, the outermost orbitals. The s-orbitals contract, the p-orbitals also contract, but the more diffuse d- and f-orbitals can become even more diffuse as electrons in the contracted s- and p-orbitals offer a greater degree of shielding to any electrons in the d- and f-orbitals.

Thallium and lead have higher values of their first ionization energies than expected from the trends down their respective groups, because their p-orbitals are more compact. The relativistic effect upon the 6p-orbitals of the elements from Tl to Rn is to reinforce a stabilization of one orbital with respect to the other two. Instead of the expected trend, the first ionization energies of Tl, Pb and Bi ($+589$, $+715$ and $+703$ kJ mol^{-1}) do not show a general increase like those of In, Sn and Sb ($+558$, $+709$ and $+834$ kJ mol^{-1}); the value for Bi is lower than that of Pb.

Another notable difference in properties down groups is the "**inert pair effect**", as demonstrated by the chemical behaviour of Tl, Pb and Bi. The main oxidation states of these elements are $+1$, $+2$ and $+3$, respectively, which are lower by two units than those expected from the behaviour of the lighter members of each group. There is a smaller, but similar, effect in the chemistry of In, Sn and Sb. These effects are partially explained by the relativistic effects on the appropriate ionization energies, which make the achievement of the higher oxidation states (the participation of the pair of s-electrons in chemical bonding) relatively more difficult.

The $+3$ state of Tl is much less stable than the $+1$ state. The reduction potential for Al^{3+} shows that the metal is a fairly good reducing agent, but it does not react with aqueous acids because the surface is normally covered with a protective oxide layer, which is impervious to ions in solution. In the form of a fine powder the metal is a powerful reducing agent. It is used as the reducing agent in the booster rockets of the Space Shuttle, in which the oxidizing agent is ammonium chlorate(VII).

The heavier metals react with dilute acids to give Ga^{3+}, In^{3+} and Tl^{+}, respectively. In alkaline solution the only simple ionic form of boron is the B(OH)$_4^-$ ion. Ions of the same form, but complicated by further hydration and polymerization, are produced when Al and Ga react with sodium hydroxide solution.

6.2.4 $E°$ Values for the Reduction of Sodium, Magnesium and Aluminium Ions to the Metallic State

Across the s- and p-blocks the reducing powers of the elements decrease. This is shown for the elements Na, Mg and Al, appropriate data being presented in Table 6.8.

Table 6.8 Data and calculated E° values for the reduction potentials of Na, Mg and Al (enthalpies in kJ mol^{-1})

	Na	Mg	Al
$\Delta_a H^\circ$ (M, g)	+108	+147	+330
I_1	+496	+738	+578
I_2		+1451	+1817
I_3			+2745
$\Delta_{hyd} H^\circ (M^{n+}, g)$	−424	−1963	−4741
E°(calc)/V	−2.49	−2.42	−1.83
E°(expt)/V	−2.71	−2.37	−1.66

The reduction enthalpies are calculated, as in the above cases, with suitable modification for the varying number of electrons involved. The general equation for the calculation of the reduction enthalpies is:

$$\text{Reduction enthalpy} = -\Delta_{hyd} H^\circ(M^{n+}, g) - \Sigma I_{1-n}(M) - (n \times 6.2)$$
$$- \Delta_a H^\circ(M, g) + (426.2 \times n)$$
$$= -\Delta_{hyd} H^\circ(M^{n+}, g) - \Sigma I_{1-n}(M)$$
$$- \Delta_a H^\circ(M) + (420 \times n) \qquad (6.10)$$

where $n = 1$ for Na, 2 for Mg and 3 for Al.

The calculations are consistent with the observed trends, and it should be noticed that a major factor responsible for the lessening reducing power of the elements is the increasing total ionization energy, which is not offset sufficiently by the relatively smaller increases in hydration enthalpy. The total ionization energies and the negative enthalpies of hydration needed for the calculation of the reduction potentials of Na, Mg and Al are plotted in Figure 6.3.

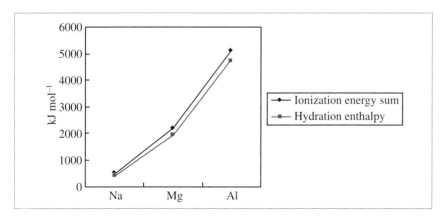

Figure 6.3 Plots of the sums of ionization energies and the negative values of the enthalpies of hydration for Na, Mg and Al

It is clear from this example and those of Section 6.2.2 that quite simple calculations can identify the major factors that govern the values of E° for any couple, and that major trends may be rationalized.

Worked Problem 6.5

Q From the data given in Table 6.8, comment on the role of the atomization enthalpies in determining the trend in the observed standard reduction potentials for the Na, Mg and Al couples.

A The difference between the ionization energy sums and the negative values of the enthalpies of hydration increases in the order Na < Mg < Al, and that this difference is the main determinant of the trend in the reduction potentials. The trend in atomization enthalpies is also in the order Na < Mg < Al, and the greater the value of the atomization enthalpy, the less negative is the value of the reduction potential for the couple.

6.2.5 The Aqueous Chemistry of the Group 14 Elements

The Group 14 elements show a relatively small amount of ionic chemistry because of the virtual impossibility of formation of ions with a charge of $+4$. Carbon dioxide dissolves in alkaline solution to give the hydrogen carbonate and carbonate ions, HCO_3^- (aq) and CO_3^{2-}. In acid solutions these ions are protonated to give carbonic acid, H_2CO_3, which decomposes with the release of carbon dioxide. The first three elements are non-metallic, and the metals Sn and Pb are subject to the inert-pair effect. The sparse data for the ions of the elements of the group are given in Table 6.9.

Group 14		
12.011	**C**	
6	$2s^2 2p^2$	2.5
28.086	**Si**	
14	$3s^2 3p^2$	1.7
72.64	**Ge**	
32	$3d^{10} 4s^2 4p^2$	2.0
118.710	**Sn**	
50	$4d^{10} 5s^2 5p^2$	1.7
207.2	**Pb**	
82	$4f^{14} 5d^{10} 6s^2 6p^2$	1.6

Table 6.9 Standard reduction potentials for the Group 14 elements at pH = 0 and pH = 14

E°/V at pH = 0					E°/V at pH = 14				
IV		II		0	IV		II		0
CO_2	-0.1	CO	-0.52	C	CO_3^{2-}			-1.15	C
					SiO_3^{2-}			-1.7	Si
					GeO_3H^-			-0.89	Ge
Sn^{4+}	$+0.15$	Sn^{2+}	-0.138	Sn	$Sn(OH)_6^{2-}$	-0.93	SnO_2H^-	-0.91	Sn
		Pb^{2+}	-0.126	Pb					

Only Sn and Pb have any ionic chemistry in acidic solution, and both elements show inert-pair behaviour, with Pb^{2+} as the more stable state for that element. Only Sn has properly characterized ionic chemistry in alkaline solution, but the +2 oxides of both Sn and Pb are amphoteric, reacting with an excess of sodium hydroxide solution to give oxoanions.

6.2.6 The Aqueous Chemistry of the Group 15 Elements

The standard reduction potentials for the elements of Group 15 at pH values of 0 and 14, respectively, are given in Tables 6.10 and 6.11, which include only the main and well-characterized ions and some molecular species.

Group 15		
14.007		
	N	
7		3.1
	$2s^2 2p^3$	
30.974		
	P	
15		2.1
	$3s^2 3p^3$	
74.922		
	As	
33		2.2
	$3d^{10} 4s^2 4p^3$	
121.760		
	Sb	
51		1.8
	$4d^{10} 5s^2 5p^3$	
208.980		
	Bi	
83		1.7
	$4f^{14} 5d^{10} 6s^2 6p^3$	

Table 6.10 Standard reduction potentials for the Group 15 elements at pH = 0

V		IV		III		II		I		0
NO_3^-	+0.8	N_2O_4	+1.07	HNO_2	+0.98	NO	+1.59	N_2O	+1.77	N_2
H_3PO_4	−0.93	$H_4P_2O_6$	+0.38	H_3PO_3			−0.5	H_3PO_2	−0.51	P
H_3AsO_4			+0.56	$HAsO_2$			+0.25			As
				SbO^+			+0.21			Sb
				BiO^+			+0.32			Bi

Table 6.11 Standard reduction potentials for the Group 15 elements at pH = 14

V		IV		III		II		I		0
NO_3^-	−0.85	N_2O_4	+0.87	NO_2^-	−0.46	NO	+0.76	N_2O	+0.94	N_2
PO_4^{3-}			−1.12	HPO_3^{2-}			−1.57	$H_2PO_2^-$	−2.05	P
AsO_4^{3-}			−0.71	AsO_2^-				−0.68			As
$Sb(OH)_6^-$			−0.465	$Sb(OH)_4^-$				−0.64			Sb

Nitrogen has some water-stable ions in which it has negative oxidation states. The Latimer diagram for these ions, the protonated forms of hydroxylamine, NH_3OH^+, and hydrazine, $N_2H_5^+$, and the ammonium ion, NH_4^+, is shown below:

0		−1		−2		−3
$\frac{1}{2}N_2$	−1.87	NH_3OH^+	+1.41	$N_2H_5^+$	+1.28	NH_4^+

A volt-equivalent diagram for the water-soluble nitrogen species in acidic solution is shown in Figure 6.4. It shows that the nitrate(V) ion is the least stable species, but also indicates the meta-stability of nitrous acid, which is unstable with respect to disproportionation into oxidation states $+5$ and zero:

$$5HNO_2(aq) + 3H_2O(l) \rightarrow 3H_3O^+(aq) + 3NO_3^-(aq) + N_2(g) \quad (6.11)$$

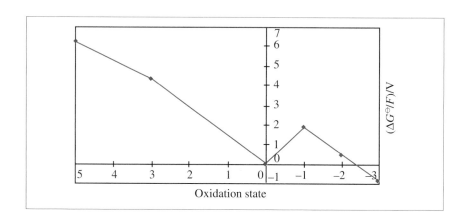

Figure 6.4 A volt-equivalent diagram for the water-soluble states of nitrogen at pH = 0

The overall standard potential for the equation is the *difference* between the potential for the reduction half-reaction ($+1.45$ V) and that for the oxidation half-reaction ($+0.94$ V): $E^\circ = 1.45 - 0.94 = +0.51$ V, which, being positive, means that the reaction is feasible.

The reduction potential for the nitrate(V)/nitrate(III) couple in acid solution of $+0.94$ V indicates from the limited data in Table 6.12 that nitrate(V) ion in acidic solution is a reasonably good oxidizing agent. However, nitric acid as an oxidant usually functions in a different manner, with the production of "brown fumes" with a metal (*e.g.* copper) or a metal sulfide (*e.g.* FeS_2). The brown fumes consist of N_2O_4 (brown gas) and its monomer NO_2 (colourless gas). Concentrated nitric acid consists of about 70% of the acid in an aqueous solution. In such a solution there is some dissociation of the nitric acid molecules to give the nitronium ion, NO_2^+, which represents the primary oxidizing species:

$$2HNO_3(conc) \rightleftharpoons NO_2^+(aq) + NO_3^-(aq) + H_2O(l) \quad (6.12)$$

Of the ions that have nitrogen in negative oxidation states, the ammonium ion is the most stable and the intermediate -1 and -2 states (protonated hydroxylamine and protonated hydrazine, respectively) are oxidizing agents and are unstable with respect to disproportionation.

The most stable state of nitrogen in acidic solution is the ammonium ion, $NH_4^+(aq)$, which is isoelectronic with CH_4 and H_3O^+. It is a tetrahedral ion with strong N–H bonds. The mean N–H bond enthalpy in $NH_4^+(aq)$ is 506 kJ mol^{-1} (that of the O–H bonds in H_3O^+ is 539 kJ mol^{-1}). The enthalpy of hydration of the ammonium ion is -345 kJ mol^{-1}. This value placed into the Born equation (3.32) gives an estimate of the radius of the ammonium ion of 135 pm, a value insignificantly different from its thermochemical radius of 136 pm. The value is comparable to that estimated for the smaller H_3O^+ ion (99 pm) from its more negative enthalpy of hydration (-420 kJ mol^{-1}, see Section 2.6.1). The **proton affinity** of the ammonia molecule is of interest in a comparison of its properties with those of the water molecule. The proton affinity is defined as the standard enthalpy change for the reaction:

$$NH_3(g) + H^+(g) \rightarrow NH_4^+(g) \qquad (6.13)$$

and has a value of -852 kJ mol^{-1}. The corresponding value for the water molecule is -690 kJ mol^{-1} (see Section 2.6.1) and is consistent with the greater basicity of the ammonia molecule. Nitrate(V) loses its powers of oxidation in alkaline solution.

The two positive oxidation states of P ($+5$ and $+3$) are both more stable than their nitrogen equivalents, and phosphoric acid has no oxidant properties apart from those given by the hydrated protons produced from it in aqueous solution. A dilute solution of phosphoric acid will provide a sufficiently high concentration of hydrated protons to oxidize any metal to its most stable state, providing the reduction potentials for the metal ion/metal couple are negative.

The corresponding As species in acid solution are less stable than those of P due to the effects of the 3d contraction. Further down the group, Sb and Bi show more metallic behaviour, with the two positive ions indicated in Table 6.11, and Sb shows amphoteric behaviour: the $+3$ oxide reacts with sodium hydroxide solution.

The estimated values of the N–H bond enthalpy, the enthalpy of hydration of the ammonium ion and the proton affinity of the ammonia molecule, given in this text, are taken from the solution to Problem 6.6.

Hypervalence in the Phosphate Ion

The phosphate ion, PO_4^{3-}, together with many oxoanions in which the notional content of the valence shell of the central atom is in excess of eight, may be described as exhibiting hypervalence, *i.e.* the central atom is surrounded by more than an octet of electrons in its valence shell. Such octets normally occupy the four molecular orbitals (MOs) that have contributions from the ns and np atomic orbitals of the central atom, where n is the highest value of the principal quantum

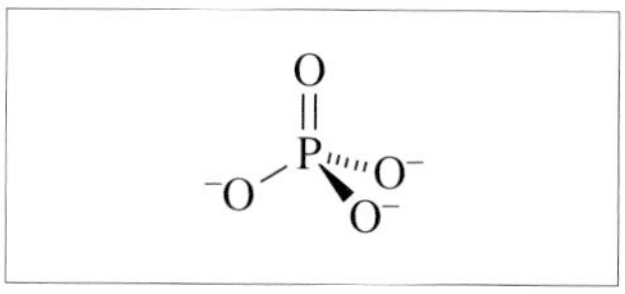

Figure 6.5 One method of representing the bonding in the phosphate ion

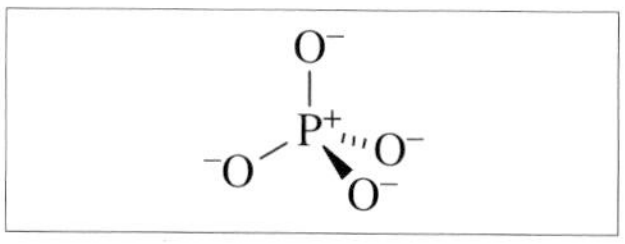

Figure 6.6 A form of the phosphate ion which does not show hypervalency of the phosphorus atom

Details of MO theory and the symmetry symbols used in the description of the bonding in PO_4^{3-} and SF_6 are given in related text *Structure and Bonding*.[1]

number relating to the valency electrons. The tetrahedral phosphate ion may be regarded as having localized bonding, as shown in Figure 6.5, in which there are three single bonds to negatively charged oxygen atoms and a double bond to the fourth oxygen atom, although that is only one of many possible canonical forms that could contribute to the overall structure.

To explain the existence of a conventional double bond in the phosphate ion requires five electrons to be supplied by the central phosphorus atom, which may enter into MO formation with suitable orbitals from the ligand oxygen atoms. To arrange this requires the use of one of the 3d orbitals of the phosphorus atom. This cannot occur in compounds of elements of the second period, where hypervalence is not observed. The availability of accessible d-orbitals has, in the past, been used as the justification of hypervalence in compounds of the subsequent periods. An alternative method of representing the structure of the PO_4^{3-} ion is shown in Figure 6.6.

In this canonical form the phosphorus atom has a positive charge and is isoelectronic with the silicon atom. It may then form four single covalent bonds to the four single-negatively charged oxygen atoms. In this form, the phosphorus atom is not showing hypervalency.

A MO approach to the problem indicates that although d-orbital participation is not required to explain the bonding in a tetrahedral oxoanion, some participation ensures its extra stability. Figure 6.7 is a simplified MO diagram for a tetrahedral oxoanion making use of the 3s and 3p orbitals of the central atom, X, and the group orbitals constructed from the 2p atomic orbitals of the four ligand oxygen atoms.

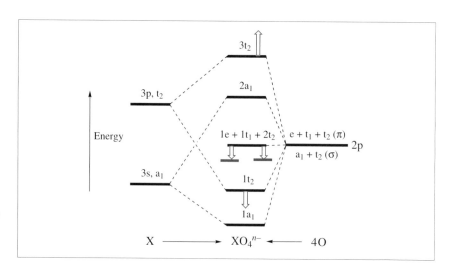

Figure 6.7 A MO diagram for tetrahedral oxoanions, XO_4^{n-}; the changes indicated in colour occur when d-orbital participation is allowed

The four pairs of 2s electrons on the oxygen atoms are omitted from the diagram for simplicity. Four 2p orbitals of the oxygen atoms may be directed towards the central X atom and form the group orbitals labelled as $a_1 + t_2$ (σ), which are σ-type orbitals. These may interact with the orbitals of the central atom that have the same symmetry, *i.e.* the 3s (a_1) and 3p (t_2) atomic orbitals, to give the bonding $1a_1$ and $1t_2$ MOs and the $2a_1$ and $3t_2$ anti-bonding MOs. The oxoanions (X = Si, P, S or Cl) possess 24 electrons that need to occupy the MOs. Two electrons occupy the $1a_1$ bonding orbital and six electrons occupy the $1t_2$ bonding orbitals. The eight electrons give a bond order of one to each of the supposed X–O bonds. The other sixteen electrons are accommodated in the non-bonding sets of MOs labelled as $e + 1t_1 + 2t_2$.

The 3d orbitals of atom X transform as $e + t_2$ in tetrahedral symmetry, and although they are higher in energy than the 3p orbitals they can and do participate in the MO scheme. The two 3d orbitals that transform as the e representation of the tetrahedral group (z^2 and $x^2 - y^2$) interact with the otherwise non-bonding $1e$ set of oxygen orbitals to make $1e$ bonding, and the anti-bonding combination is labelled $2e$. The other three 3d orbitals of atom X transform as t_2, and have the effect of stabilizing the $1t_2$ level. This increases its bonding effect. The $2t_2$ level is also slightly stabilized. The $3t_2$ anti-bonding orbital becomes more anti-bonding, and the highest energy level is that labelled $4t_2$, which is also anti-bonding. If such 3d participation occurs the bond orders of the X–O bonds are increased to a value greater than one and the ion as a whole is stabilized as a result. The extent of this stabilization depends upon the 3p–3d energy gap.

Experimental evidence for the occurrence of hypervalence is given by studies of the vibrational spectra of ions. Tetrahedral 5-atom ions possess $3 \times 5 - 6 = 9$ fundamental vibrational modes. They correspond to the four vibrations shown in Figure 6.8.

The symmetric stretching vibration, labelled v_1 (a_1), in which the four bonds expand and contract synchronously is a feature in the Raman spectra of the ions. The bending vibrations labelled v_2 (e) are doubly degenerate and occur at a particular frequency in the Raman spectra of the ions. There are two modes of vibration, labelled v_3 and v_4, which belong to the representation t_2. Both are triply degenerate and are represented in Figure 6.8 as mainly bond stretching (v_3) or bending (v_4) vibrations. Since they both have t_2 symmetry, they can interact so that both have bond stretching and bending components. The two separate frequencies for any tetrahedral ion are found in the Raman and infrared spectra of their compounds. Table 6.12 gives the frequencies for the four vibration modes for some tetrahedral ions.

The 24 electrons in the XO_4^{n-} ions are the four sets of $2p^4$ electrons of the ligand oxygen atoms plus the valency electrons of the central X atoms and those representing the negative charge.

In general, when extra orbital interaction of this kind occurs the occupied orbitals are stabilized at the expense of the vacant orbitals. In the case of the tetrahedral XO_4^{n-} ions, the filled $1t_2$ and $2t_2$ orbitals are stabilized and the $3t_2$ and $4t_2$ orbitals are destabilized, giving an energetic advantage to the stability of the compound. Reed and Weinhold[2] describe in detail the MOs of SF_6, one of the classical examples of hypervalence. They show that the structure is incorrectly described by the use of d^2sp^3 hybridization and that such a model overstates the influence of the 3d orbitals of the sulfur atom, although it is still used in the electron-counting procedures of the valence shell electron pair repulsion (VSEPR) theory. Nevertheless, the 3d orbital participation confers -1046 kJ mol^{-1} to the standard enthalpy of formation of the SF_6 molecule and is a major contribution to the experimental value of -1221 kJ mol^{-1}.

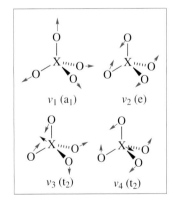

Figure 6.8 The fundamental vibrational modes of tetrahedral species

Table 6.12 Fundamental vibrational frequencies for some tetrahedral ions; the frequencies are given as wavenumbers with units of cm^{-1}

Ion	v_1	v_2	v_3	v_4
SiO_4^{4-}	819	340	956	527
PO_4^{3-}	938	420	1017	567
SO_4^{2-}	983	450	1105	611
ClO_4^{-}	928	459	1119	625

Bond lengths are a good indication of bond strength, the two having an inverse relationship. The XO distances in XO_4^{n-} ions are given in Table 6.13.

Table 6.13 X—O distances in XO_4^{n-} ions

X	X–O/pm
Si	160
P	155
S	151
Cl	148

The distances decrease in the order Si > P > S > Cl, consistent with the radii of the atoms, but are considerably smaller than the sums of the covalent radii of the atoms participating in the bonding by 25–31 pm, indicating bond orders greater than one.

There is a general increase of wavenumber in the order $SiO_4^{4-} < PO_4^{3-} < SO_4^{2-} < ClO_4^{-}$; this is consistent with increasing bond order as would be expected if hypervalence with d-orbital contributions were occurring. The trend in the values of v_2 are perhaps most indicative of the participation of π-bonding, since the bending mode would be most likely to weaken this kind of bonding.

6.2.7 The Aqueous Chemistry of the Group 16 Elements

The standard reduction potentials for the main ionic species of Group 16 are given in Tables 6.14 and 6.15 at pH values of 0 and 14, respectively. The potential for the half-reaction:

$$\tfrac{1}{2}O_2(g) + 2H^+(aq) + 2e^- \rightleftharpoons H_2O(l) \tag{6.14}$$

is the mean of the two potentials given in Table 6.14, *i.e.* $\tfrac{1}{2}(0.695 + 1.763) = +1.229$ V. This oxygen reduction potential and its values in the pH range 0–14 are very important, as discussed in Chapter 5, as they determine which ions with oxidizing powers may exist in aqueous solutions and which should oxidize water to dioxygen.

Table 6.14 Standard reduction potentials for the Group 16 elements at pH = 0

E°/V at pH = 0							
VI		IV		0		-1		-2
				$\tfrac{1}{2}O_2$	0.695	$\tfrac{1}{2}H_2O_2$	1.763	$\tfrac{1}{2}H_2O$
HSO_4^-	0.158	H_2SO_3	0.45	S		0.14		H_2S
SeO_4^{2-}	1.15	H_2SeO_3	0.74	Se		-0.44		H_2Se
H_6TeO_6			0.74	Te		-0.79		H_2Te

Table 6.15 Standard reduction potentials for the Group 16 elements at pH = 14

E°/V at pH = 14

VI		IV		0		−1		−2	
				$\frac{1}{2}O_2$	−0.065	$\frac{1}{2}HO_2^-$	0.87	OH^-	
SO_4^{2-}	−0.93	SO_3^{2-}	−0.66	S			−0.48	HS^-	
SeO_4^{2-}	0.05	H_2SeO_3	−0.36	Se			−0.67	Se^{2-}	
TeO_4^{2-}	0.07	TeO_3^{2-}	−0.57	Te			−1.14	Te^{2-}	

Worked Problem 6.6

Q Comment on the stability of hydrogen peroxide at pH = 0.

A The H_2O_2/H_2O couple has a high positive reduction potential (Table 6.14) and because the O_2/H_2O_2 potential is considerably smaller, hydrogen peroxide is unstable with respect to disproportionation:

$$H_2O_2(aq) + 2H^+(aq) + 2e^- \rightleftharpoons 2H_2O(l) \qquad E^{\circ} = +1.763 \text{ V}$$

$$O_2(aq) + 2H^+(aq) + 2e^- \rightleftharpoons H_2O_2(aq) \quad E^{\circ} = +0.695 \text{ V}$$

The overall reaction produced by reversing the second half-reaction and adding it to the first is:

$$2H_2O_2(aq) \rightleftharpoons O_2(g) + 2H_2O(l) \quad E^{\circ} = +1.763 - 0.695 = +1.068 \text{ V}$$

Group 16		
15.999		
	O	
8		3.5
	$2s^2 2p^4$	
32.065		
	S	
16		2.4
	$3s^2 3p^4$	
78.96		
	Se	
34		2.5
	$3d^{10} 4s^2 4p^4$	
127.60		
	Te	
52		2.0
	$4d^{10} 5s^2 5p^4$	
208.982^a		
	Po	
84		1.8
	$4f^{14} 5d^{10} 6s^2 6p^4$	

^{a}Most stable isotope

Solutions of hydrogen peroxide decompose slowly and are best kept at low temperatures. Some solid compounds of Po, *e.g.* $PoCl_2$, indicate that it has metallic properties, but most of its chemistry is that of metal polonides containing the Po^{2-} ion.

From Group 15 to Group 16, non-metallic behaviour takes over completely with no positive ions being stable. The +6 state of sulfur is seen to have very poor oxidizing properties, and it is only in its concentrated form, and when hot, that sulfuric(VI) acid is a good oxidant. Hot concentrated sulfuric acid oxidizes metallic copper and is reduced to sulfur dioxide. The relative stabilities of the Se species with positive oxidation states are considerably less than their S or Te counterparts, another example of the effect of the 3d contraction.

The stabilities of the dihydrides of the Group 16 elements decrease considerably down the group, from the very stable water to H_2Te, which is readily oxidized.

In alkaline solution there is a general stabilization of the positive oxidation states of the elements of the group and a destabilization of the negative oxidation states.

18.998		
	F	
9		4.1
	$2s^2 2p^5$	

35.453		
	Cl	
17		2.8
	$3s^2 3p^5$	

79.904		
	Br	
35		2.7
	$3d^{10} 4s^2 4p^5$	

126.904		
	I	
53		2.2
	$4d^{10} 5s^2 5p^5$	

209.987^a		
	At	
85		2.0
	$4f^{14} 5d^{10} 6s^2 6p^5$	

Group 17

^{a}Most stable isotope

6.2.8 The Aqueous Chemistry of the Group 17 Elements

The standard reduction potentials for the main species formed by the Group 17 elements in aqueous solution are given in Tables 6.16 and 6.17, for pH values 0 and 14, respectively. Irrespective of the pH of the solution, the halogen elements range from the extremely powerful F_2 (which has the potential to oxidize water to dioxygen), through the powerful oxidants Cl_2 and Br_2, to I_2, which is a relatively weak oxidant.

Table 6.16 Standard reduction potentials for the Group 17 elements at pH = 0

E°/V at pH = 0

VII		V		III		I		0		−1
								$1/2F_2$	+3.05	HF
ClO_4^-	+1.19	ClO_3^-	+1.21	$HClO_2$	+1.65	HOCl	+1.61	$1/2Cl_2$	+1.36	Cl^-
BrO_4^-	+1.85	BrO_3^-			+1.45		HOBr	+1.60	$1/2Br_2$	+1.09	Br^-
H_5IO_6	+1.6	IO_3^-			+1.13	IO^-	+1.44	$1/2I_2$	+0.54	I^-

All the positive oxidation states of the Group 17 elements are powerful oxidants. In alkaline conditions the positive oxidation states are still reasonably powerful oxidants, *e.g.* HOCl is the basis of some household bleach solutions which allegedly kill all known germs. Dilute chloric(VII) acid has a high reduction potential, but reacts very slowly with most reducing agents.

Bromine(VII) shows even more thermodynamic instability than its Cl and I neighbours; this is another example of the alternation of properties down the p-block groups brought about by the 3d contraction.

Table 6.17 Standard reduction potentials for the Group 17 elements at pH =14

E°/V at pH = 14

VII		V		III		I		0		−1
								$1/2F_2$	+2.87	F^-
ClO_4^-	+0.37	ClO_3^-	+0.30	ClO_2^-	+0.68	ClO^-	+0.42	$1/2Cl_2$	+1.36	Cl^-
BrO_4^-	+1.03	BrO_3^-			0.49	BrO^-	+0.46	$1/2Br_2$	+1.09	Br^-
H_5IO_6	+0.65	IO_3^-			0.15	IO^-	+0.42	$1/2I_2$	+0.54	I^-

6.2.9 The Aqueous Chemistry of Xenon

Table 6.18 contains the small amount of data for the + 6 and + 7 states of Xe. All four species are extremely unstable and must be handled very carefully. It is probable that radon has a similar chemistry, and there is a possibility that some higher oxidation states of krypton exist.

Table 6.18 Standard reduction potentials for xenon at pH = 0 and pH = 14

$E°/V$ at pH = 0					$E°/V$ at pH = 14				
VIII		VI		0	VIII		VI		0
H_4XeO_6	2.42	XeO_3	2.1	Xe	$HXeO_8^-$	0.94	$HXeO_4^-$	1.26	Xe

6.3   Summary of the s- and p-Block Periodicity

As is the case in the solid state fluorides and oxides, [3] there is a transition from purely metallic behaviour in the Group 1 (apart from H) and Group 2 elements, which exist in aqueous solutions as positive ions, through Groups 13, 14 and 15, where metallic character decreases across the periods and increases down each group, to the elements of Groups 16, 17 and 18 which are almost entirely non-metallic. Possible exceptions are in the chemistry of Po and At, which are not well characterized and so are omitted from the discussions. The behaviour of the elements changes from the basic-oxide-forming elements of the first two groups to the acidic behaviour of the positive oxidation states of the later p-block elements. Intermediate groups show the transition from non-metallic acidic behaviour, through some amphoteric oxidation states, to the basic metallic properties of the lower oxidation states of the heavier members of those groups.

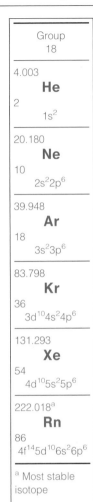

Group 18		
4.003		
He		
2		
$1s^2$		
20.180		
Ne		
10		
$2s^2 2p^6$		
39.948		
Ar		
18		
$3s^2 3p^6$		
83.798		
Kr		
36		
$3d^{10}4s^2 4p^6$		
131.293		
Xe		
54		
$4d^{10}5s^2 5p^6$		
222.018^a		
Rn		
86		
$4f^{14}5d^{10}6s^2 6p^6$		

[a] Most stable isotope

Summary of Key Points

1. The terms valency and oxidation state were defined and exemplified by reference to the s- and p-block elements.

2. The periodicity in the redox behaviour of the s- and p-block elements was described. Group-by-Group discussions were included.

3. Variations in the standard reduction potentials of the Group 1 M^+/M couples were interpreted in terms of a thermochemical cycle.

4. The trend in reducing power of the metals Na, Mg and Al was explained.

5. Valencies and oxidation states that vary from those expected from the operation of the octet rule were explained; hypervalency and the inert pair effect were described.

References

1. J. Barrett, *Structure and Bonding*, RSC Tutorial Chemistry Text, no. 5, Royal Society of Chemistry, Cambridge, 2001.
2. A. E. Reed and F. Weinhold, *On the Role of d Orbitals in SF$_6$*, in *J. Am. Chem. Soc.*, 1986, **108**, 3586.
3. J. Barrett, *Atomic Structure and Periodicity*, RSC Tutorial Chemistry Text, no. 9, Royal Society of Chemistry, Cambridge, 2002.

Further Reading

W. M. Latimer, *Oxidation Potentials*, 2nd edn., Prentice-Hall, New York, 1952. As its title implies, this book contains oxidation potentials (*i.e.* reduction potentials with their signs reversed, as was the convention at the time of its publication) for all the elements. This is a classic book, full of good chemistry, written by the inventor of Latimer diagrams and still available in libraries.

J. Emsley, *The Elements*, 3rd edn., Oxford University Press, Oxford, 1999. This compilation of data contains detailed Latimer diagrams for all the elements that have ionic species which exist in aqueous solution.

F. A. Cotton, G. Wilkinson, C. A. Murillo and M. Bochmann, *Advanced Inorganic Chemistry*, 6th edn., Wiley, New York, 1999.

Problems

6.1. From your understanding of the inert-pair effect and the redox properties of Tl and I$_2$, consider the apparent oxidation state of Tl in the compound TlI$_3$ and indicate what the realistic value is. The standard reduction potential for the Tl^{III}/Tl^{I} couple is +1.25 V.

6.2. The Group 17 elements are the most electronegative of their respective periods, and the reductions of the elements to their uninegative ions are thermodynamically feasible, as can be seen from the values for E° given in the table below (enthalpies in kJ mol^{-1}). Calculate enthalpy-only values for the reduction potentials for the reduction of the elements to their uninegative ions.

	F	Cl	Br	I
$\Delta_a H°$ (X, g)	+79	+121	+112	+107
$E_{ea}(X)$	− 328	− 349	− 325	− 295
$\Delta_{hyd}lH°(X^-, g)$	− 504	− 359	− 328	− 287
$E°/V$	+2.87	+1.36	+1.09	+0.54

6.3. Write an equation for the reaction between Br^V and Br^- at pH = 0. From the data in Table 6.16, decide what the products are, and calculate the overall potential of the reaction.

6.4. At pH = 0, which halogen in its +6 state is the most powerful oxidant with respect to its reduction to the − 1 state?

6.5. From the data (kJ mol^{-1}) in the table, calculate (i) a value for the enthalpy of hydration of the ammonium ion, (ii) a mean value for the N–H bond enthalpy in the ammonium ion, (iii) a value for the proton affinity of the ammonia molecule [*i.e.* the value of $\Delta_{prot}H°$ for the reaction $NH_3(g) + H^+(g) \rightarrow NH_4^+(g)$], and (iv) a mean value for the strength of the N–H bonds in the ammonium ion.

	$\Delta_f H°$	$\Delta_{latt}H°$	$\Delta_{sol}H°$	$\Delta_{hyd}H°$
N(g)	+473			
$NH_4ClO_4(s)$	− 295.3	− 586.0		
$(NH_4)_2SO_4(s)$	− 1180.9	− 1789		
$NH_3(g)$	− 46.0			
NH_4^+			− 132.5	
ClO_4^-			− 129.3	− 205
SO_4^{2-}			− 909.3	− 1099

7

Periodicity of Aqueous Chemistry II: d-Block Chemistry

This chapter consists of a description of the ions formed in aqueous solutions by the transition elements – the d-block elements – and a discussion of the variations of their redox properties across the Periodic Table from Group 3 to Group 12. There is particular emphasis on the first transition series from scandium to zinc in the fourth period, with summaries of the solution chemistry of the second (Y to Cd) and third (Lu to Hg) series. The d-block ions in solution are those restricted solely to aqua complexes of cations, e.g. $[Fe(H_2O)_6]^{2+}$, and the various oxocations and oxoanions formed, e.g. VO^{2+} and MnO_4^-. Oxidation states that are not well characterized are omitted or referred to as such.

By the end of this chapter you should understand:

State symbols are omitted in this chapter, where they would be fairly obvious.

- The variations of ionic forms of transition element ions
- The derivations of enthalpies of hydration of the transition element ions
- The variations in redox behaviour across the first transition series
- The summaries of the redox chemistry of each Group of transition elements

7.1   The Ions Formed by the Transition Elements

The classical cases of distinction between valency and oxidation state occur in the **coordination complexes** of the transition elements. For example, in the **complex compound** $[Cr(NH_3)_6^{3+}](Cl^-)_3$ the **complex ion** containing the chromium ion, $[Cr(NH_3)_6]^{3+}$, has a chromium atom at its centre which engages in six electron-pair bonds with the six ammonia **ligand** molecules, the latter supplying two electrons each in the formation of six **coordinate or dative bonds.** To that extent the chromium atom is participating in six electron-pair bonds and could be called six-valent. Because the chromium atom is using none of its valency electrons, to think of it as exerting any valency is misleading. What is certain is that the complex ion exists as a separate entity, with three chloride ions as counter ions, and therefore has an overall charge of $+3$. Since the ammonia ligands are neutral molecules, the chromium atom may be thought of as losing three of its valency electrons to give the $+3$ oxidation state, written in the form Cr^{III}. The original electronic configuration of the chromium atom's valency shell is $4s^1 3d^5$, which in its $+3$ state would become $3d^3$. The complex ion is fully described as the hexaamminechromium(III) ion. Table 7.1 summarizes the oxidation states of the transition elements that are stable in aqueous solution in the absence of any ligand other than the water molecule.

Table 7.1 Oxidation states of the transition elements in acidic aqueous solution; those shown in red are the most stable in aqueous solution in the absence of dioxygen

First transition period (3d):

Sc	Ti	V	Cr	Mn	Fe	Co	Ni	Cu	Zn
				+7					
			+6	+6	+6				
		+5		+5					
	+4	+4		+4	+4				
+3	+3	+3	+3	+3	+3	+3			
		+2	+2	+2	+2	+2	+2	+2	+2
								+1	

Second transition period (4d):

Y	Zr	Nb	Mo	Tc	Ru	Rh	Pd	Ag	Cd
					+8				
				+7	+7				
			+6		+6				
		+5	+5		+5				
	+4		+4		+4				
+3			+3		+3	+3			
							+2		+2
								+1	+1

Third transition period (5d):

Lu	Hf	Ta	W	Re	Os	Ir	Pt	Au	Hg
					+8				
				+7					
			+6						
		+5							
	+4								
+3						+3			
									+2
								+1	+1
3	4	5	6	7	8	9	10	11	12

Group

The chemistry of the elements of the fourth transition period is not included in this text. It is not very well developed as yet, but follows the trends in the third period elements as might be expected. The totally synthetic elements of the fourth transition period are named as lawrencium (Lr, 103), rutherfordium (Rf, 104), dubnium (Db, 105), seaborgium (Sb, 106), bhorium (Bh, 107), hassium (Hs, 108), meitnerium (Mt, 109) and darmstadtium (Ds, 110). Other, as yet nameless, elements whose syntheses have been reported have atomic numbers 110, 111, 112, 114, 116 and 118, although some of these have not been verified.

Cation hydrolysis in the s- and p-block elements does occur, but usually at much higher pH values than those affecting the transition element cations. Thus, the sodium ion hydrolyses at a pH above 14.8, Mg^{2+} at pH values around 11.4, and Al^{3+} at pH values above 5. The influence of ionic size (Na^+, 102 pm radius; Mg^{2+}, 72 pm; Al^{3+} 54 pm) and charge is clear from these few data. For similar sizes and charges, transition metal ions undergo hydrolysis at much lower values of pH. The reason for this is their capacity to enter into coordinate bonding with ligands.

The μ-OH terminology indicates that the two OH groups that bridge the two Fe^{III} atoms are the oxygen atoms bonded to both iron atoms.

Table 7.2 Examples of first-row transition element aqua cations in aqueous solution

+2 Ion	+3 Ion
	Sc^{3+}
	Ti^{3+}
V^{2+}	V^{3+}
Cr^{2+}	Cr^{3+}
Mn^{2+}	Mn^{3+}
Fe^{2+}	Fe^{3+}
Co^{2+}	Co^{3+}
Ni^{2+}	
Cu^{2+}	
Zn^{2+}	

All the ions in Table 7.2, except Cr^{2+} and Cu^{2+}, are normally regarded as regular hexaaqua complexes. The exceptions may have six-coordination of water molecules, but with two at slightly longer distances from the central ion than the other four.

For reasons explained in Section 7.3, the hydrated copper(II) ion is formulated as $[Cu(H_2O)_4]^{2+}$, but in solution it has two further water molecules close to the central ion at a slightly larger distance away than the four coordinated water molecules. The hydrated Cr^{2+} probably is similarly irregularly coordinated.

In aqueous solutions, in which the most probable ligand is the water molecule, most of the lower oxidation states (*i.e.* +2, +3 and some of the +4 states) of transition metal ions are best regarded as hexaaqua complex ions, *e.g.* $[Fe^{II}(H_2O)_6]^{2+}$. In these ions the six coordinated water molecules are those that constitute the first hydration sphere, and it is normally accepted that such ions would have a secondary hydration sphere of water molecules that would be electrostatically attracted to the positive central ion. The following discussion includes only the aqua cations that do not, at pH = 0, undergo hydrolysis. For example, the iron(III) ion is considered quite correctly as $[Fe(H_2O)_6]^{3+}$, but at pH values higher than 1.8 the ion participates in several hydrolysis reactions, which lead to the formation of polymers and the eventual precipitation of the iron(III) as an insoluble compound as the pH value increases, *e.g.*:

$$[Fe(H_2O)_6]^{3+} \rightleftharpoons [Fe(H_2O)_5OH]^{2+} + H^+ \tag{7.1}$$

$$[Fe(H_2O)_5OH]^{2+} \rightleftharpoons [Fe(H_2O)_4(OH)_2]^+ + H^+ \tag{7.2}$$

$$2[Fe(H_2O)_5OH]^{2+} \rightleftharpoons [(H_2O)_4Fe(\mu-OH)_2Fe(H_2O)_4]^{4+} + 2H_2O \tag{7.3}$$

$$\cdots \rightleftharpoons FeO(OH) \text{ [one form of } Fe_2O_3 \cdot xH_2O, \text{ where } x = 1] \tag{7.4}$$

Examples of aqua cations containing transition elements are given in Table 7.2. The ions in Table 7.2 are those that are thermodynamically stable in acidic aqueous solutions except for Co^{3+}, which oxidizes water slowly. In alkaline solutions they are precipitated as hydroxides, oxides or hydrated oxides.

Worked Problem 7.1

Q Write an equation to show a +2 transition metal hydroxide dissolving in an acidic solution to give the hydrated +2 ion.

A Taking iron(II) hydroxide as the example of a +2 transition metal hydroxide, the equation:

$$Fe(OH)_2(s) + 2H^+(aq) \rightleftharpoons Fe^{2+}(aq) + 2H_2O(l) \tag{7.5}$$

indicates the solid compound dissolving to give the hydrated iron(II) ion.

The higher oxidation states of the transition elements may be considered to be hydrolysis products of hypothetical more highly charged cations in which the central metal ion is sufficiently electronegative to be able to participate in covalent bonding. For example, the hypothetical Mn^{7+} ion interacts with water to give an oxoanion, the manganate(VII) ion:

$$\text{"}Mn^{7+}\text{"} + 4H_2O(l) \rightarrow MnO_4^-(aq) + 8H^+(aq) \tag{7.6}$$

Examples of oxidation states of the transition elements that exist in aqueous solutions as oxocations or oxoanions are given in Table 7.3. Oxoanions tend to predominate over oxocations as the oxidation state of the central metal ion increases.

Table 7.3   Examples of oxocations and oxoanions of the transition elements in acidic and alkaline solutions

Oxidation state	Ionic form in acidic solution	Ionic form in alkaline solution
Ti(IV)	TiO^{2+}	–
V(V)	VO_2^+ or $V(OH)_4^+$	VO_4^{3-}
Cr(VI)	$Cr_2O_7^{2-}$	CrO_4^{2-}
Mn(VII)	MnO_4^-	MnO_4^-

Worked Problem 7.2

Q   Write an equation that explains the difference in forms of the V^V ion in acidic and alkaline solutions.

A   The equation for the equilibrium:

$$VO_2^+(aq) + 4OH^-(aq) \rightleftharpoons VO_4^{3-}(aq) + 2H_2O(l) \qquad (7.7)$$

indicates that the oxocation reacts with aqueous hydroxide ion to give the oxoanion, the latter being the more stable form in alkaline solution. The equilibrium would be shifted to the left if the solution were to become acidic, and the equilibrium would then best be described by the equation:

$$VO_4^{3-}(aq) + 4H_3O^+(aq) \rightleftharpoons VO_2^+(aq) + 6H_2O(l) \qquad (7.8)$$

The equation in the worked problem is very much simplified since there are several intermediate polymeric anions formed as the solution is made more acidic. The first stage is considered to be:

$$10VO_2^+(aq) + 12OH^-(aq) \rightleftharpoons V_{10}O_{24}(OH)_4^{2-}(aq) + 4H_2O(l) \qquad (7.9)$$

Intermediate oxidation states of some transition metal oxides are insoluble in water because their lattice enthalpies are sufficiently large to cause the solution of the compounds in water to give ions to be endothermic. Examples are V_2O_3, MoO_2, WO_2, MnO_2 and ReO_2.

The transition from positive ions with low oxidation states, *via* insoluble oxides with intermediate oxidation states, to oxoanions with high oxidation states, is caused by the competition between ionization energies, lattice enthalpies and enthalpies of hydration, similar to the discussion of the variations of ionic forms of the p-block elements given in Section 6.1. Further discussion occurs in Section 7.5.3.

7.2   Enthalpies of Hydration of Some Transition Element Cations

The basis of the estimations of the absolute enthalpies of hydration of the main group ions is dealt with extensively in Chapter 2. In this section, the same principles are applied to the estimation of the enthalpies of hydration of the monatomic cations of the transition elements, *i.e.* those of the ions M^{n+}. The standard enthalpies of formation of the aqueous ions are known from experimental measurements and their values, combined with the appropriate number of moles of dihydrogen oxidations to hydrated protons, gives the conventional values for the enthalpies of hydration of the ions concerned. Table 7.4 contains the Gibbs energies of formation and the enthalpies of formation of some ions formed by the first-row transition elements, and includes those formed by Ag, Cd, Hg and Ga.

Table 7.4 Standard Gibbs energies of formation and standard enthalpies of formation of some transition element monatomic cations and that of Ga^{3+} at 298 K (in kJ mol^{-1})

Ion	$\Delta_f G°$	$\Delta_f H°$	Ion	$\Delta_f G°$	$\Delta_f H°$
			Sc^{3+}	− 586.6	− 614.2
			Ti^{3+}	− 350.2	
V^{2+}	− 228.9		V^{3+}	− 253.6	
Cr^{2+}	− 165	− 143.5	Cr^{3+}	− 205	− 256
Mn^{2+}	− 228.1	− 220.8	Mn^{3+}	− 83.0	
Fe^{2+}	− 78.9	− 89.1	Fe^{3+}	− 4.7	− 48.5
Co^{2+}	− 54.4	− 58.2	Co^{3+}	+ 134.0	+ 92.0
Ni^{2+}	− 45.6	− 54	Cu^+	+ 50.0	+ 71.7
Cu^{2+}	+ 65.5	+ 64.8	Ag^+	+ 77.1	+ 105.6
Zn^{2+}	− 147.1	− 153.9	Hg^{2+}	+ 164.4	+ 170.2
Cd^{2+}	− 77.6	− 75.9	Ga^{3+}	− 159.0	− 211.3

Table 7.5 Estimated values of the standard enthalpies of hydration of some +2 and +3 transition metal ions and that for Ga^{3+}

Ion	$\Delta_{hyd} H°$
V^{2+}	− 1980
Cr^{2+}	− 1945
Mn^{2+}	− 1888
Fe^{2+}	− 1988
Co^{2+}	− 2051
Ni^{2+}	− 2134
Cu^{2+}	− 2135
Zn^{2+}	− 2083
Cd^{2+}	− 1847
Hg^{2+}	− 1868
Sc^{3+}	− 3989
Ti^{3+}	− 4296
V^{3+}	− 4482
Cr^{3+}	− 4624
Mn^{3+}	− 4590
Fe^{3+}	− 4486
Co^{3+}	− 4713
Ga^{3+}	− 4745

Inclusion of the absolute value of the standard enthalpy of hydration of the proton, $\Delta_{hyd} H°(H^+, g) = -1110$ kJ mol^{-1} (derived in Chapter 2), gives the absolute values for the enthalpies of hydration of the transition metal ions. The estimated values are given in Table 7.5.

Worked Problem 7.3

Q The standard enthalpy of formation of the Fe^{2+}(aq) cation is − 89.1 kJ mol^{-1}, the standard enthalpy of formation of gaseous iron atoms is + 416 kJ mol^{-1} and the first two ionization energies of iron are 762 and 1561 kJ mol^{-1}. Calculate a value for the absolute standard enthalpy of hydration of the Fe^{2+} ion, using data for the proton reduction half-reaction given in Chapter 2.

A A thermochemical cycle that may be used to give the estimate of the absolute standard enthalpy of hydration is shown in Figure 7.1.

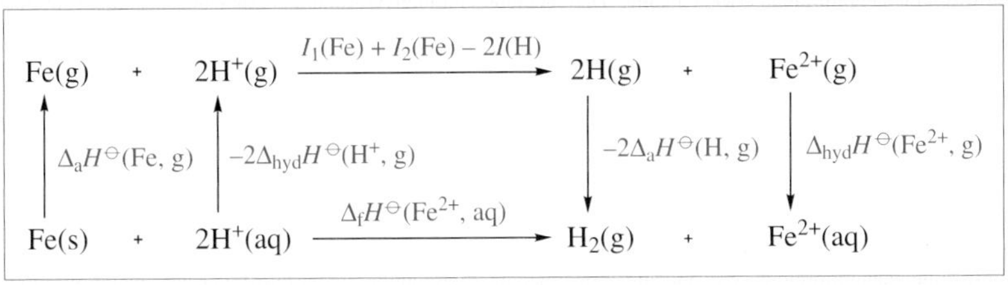

Figure 7.1 A thermochemical cycle for the calculation of the enthalpy of hydration of the Fe^{2+} ion

The estimate of $\Delta_{hyd}H^\circ(Fe^{2+}, g)$ is given by the equation:

$$\Delta_{hyd}H^\circ(Fe^{2+}, g) = \Delta_f H^\circ(Fe^{2+}, aq) - \Delta_a H^\circ(Fe, g)$$
$$+ 2\Delta_{hyd}H^\circ(H^+, g) + 2I_H - I_1(Fe)$$
$$- I_2(Fe) + 2\Delta_a H^\circ(H, g) \qquad (7.10)$$

$\Delta_{hyd}H^\circ(Fe^{2+}, g) = -89.1 - 416 - (2 \times 1110) + (2 \times 1312) - 762 - 1561 + (2 \times 218) = -1988$ kJ mol^{-1} (rounding off to the nearest 1 kJ mol^{-1})

The values of absolute enthalpies of hydration given in Table 7.5 for the $+2$ ions from V to Zn are those estimated in the same manner as that for Fe^{2+} in Worked Problem 7.3. The accepted values for the ionic radii of the $+2$ ions in Table 7.5 are given in Table 7.6, which also includes the value for the Ca^{2+} ion for purposes of comparison.

By the same method as that used in Chapter 2, the Born equation (2.43) may be used to calculate the effective radii of the ions from Cr^{2+} to Zn^{2+} and the resulting radii are found to be, on average, 62 ± 2 pm larger than the ionic radii of those ions, which is an apparent enlargement of about 82%.

The same methods are used to estimate the enthalpies of hydration of the $+3$ ions of Table 7.5. The appropriate ionic radii of the ions are given in Table 7.7. The radius of the Ga^{3+} ion (3d^{10}) is included, as the ion comes at the end of the series in which the 3d orbitals are increasingly occupied.

Table 7.6 Ionic radii of some $+2$ transition metal ions and calcium

Ion	3d	r/pm
Ca^{2+}	0	100
Ti^{2+}	2	86
V^{2+}	3	79
Cr^{2+}	4	80
Mn^{2+}	5	83
Fe^{2+}	6	78
Co^{2+}	7	75
Ni^{2+}	8	69
Cu^{2+}	9	73
Zn^{2+}	10	74

The ionic radius of Ti^{2+} is included in Table 7.6, but it is not used in the estimations of enthalpies of hydration because its presence in aqueous solution is dubious.

Worked Problem 7.4

Q The standard enthalpy of formation of the Cr^{3+}(aq) cation is -256 kJ mol^{-1}, the standard enthalpy of formation of gaseous

Table 7.7 Ionic radii of some +3 transition metal ions and that of Ga^{3+}

Ion	3d	r/pm
Sc^{3+}	0	75
Ti^{3+}	1	67
V^{3+}	2	64
Cr^{3+}	3	62
Mn^{3+}	4	65
Fe^{3+}	5	65
Co^{3+}	6	55
Ga^{3+}	10	62

Figure 7.2 A thermochemical cycle for the calculation of the enthalpy of hydration of the Cr^{3+} ion

chromium atoms is $+397$ kJ mol^{-1} and the first three ionization energies of chromium are 653, 1591 and 2987 kJ mol^{-1}. Calculate a value for the absolute standard enthalpy of hydration of the Cr^{3+} ion, using data for the proton reduction half-reaction given in Chapter 2.

A The thermochemical cycle that may be used for the estimate of the absolute standard enthalpy of hydration is shown in Figure 7.2. The estimate of $\Delta_{hyd}H^{\circ}(Cr^{3+}, g)$ is given by the equation:

$$Cr(g) + 3H^+(g) \xrightarrow{I_1(Cr) + I_2(Cr) + I_3(Cr) - 3I(H)} 3H(g) + Cr^{3+}(g)$$

$$\Delta_a H^{\ominus}(Cr, g) \uparrow \quad -3\Delta_{hyd}H^{\ominus}(H^+, g) \qquad -3\Delta_a H^{\ominus}(H, g) \downarrow \quad \Delta_{hyd}H^{\ominus}(Cr^{3+}, g) \downarrow$$

$$Cr(s) + 3H^+(aq) \xrightarrow{\Delta_f H^{\ominus}(Cr^{3+}, aq)} {}^3/_2 H_2(g) + Cr^{3+}(aq)$$

$$\Delta_{hyd}H^{\circ}(Cr^{3+}, g) = \Delta_f H^{\circ}(Cr^{3+}, aq) - \Delta_a H^{\circ}(Cr, g)$$
$$+ 3\Delta_{hyd}H^{\circ}(H^+, g) + 3I_H - I_1(Cr)$$
$$- I_2(Cr) - I_3(Cr) + 3\Delta_a H^{\circ}(H, g) \qquad (7.11)$$

$$\Delta_{hyd}H^{\circ}(Cr^{3+}, g) = -256 - 397 - (3 \times 1110) + (3 \times 312) - 653 - 1591$$
$$-2987 + (3 \times 218) = -4624 \text{ kJ mol}^{-1} \text{ (to the nearest 1 kJ mol}^{-1})$$

The ionic radii of the $+2$ and $+3$ ions given in Tables 7.6 and 7.7 are plotted against the number of 3d electrons they possess in Figure 7.3.

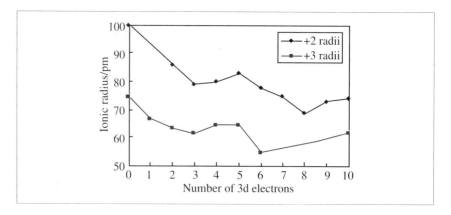

Figure 7.3 Plots of the $+2$ and $+3$ ionic radii against the number of 3d electrons the ions possess

If the Born equation (2.43) is used to calculate the effective radii of the ions from Sc^{3+} to Co^{3+}, the resulting radii are found to be, on average, 77 ± 3 pm larger than the ionic radii of those ions, an apparent enlargement of about 120%. The apparent enlargements of the $+2$ and $+3$ ions could also be due to a reduction in their apparent charges, or both effects of size and charge could be operating, indicating some delocalization of the charge on the metal ion to include at least the primary hydration sphere.

7.3   Variations in the Enthalpies of Hydration of the $+2$ and $+3$ Ions of the Transition Elements

Across the sequence of elements, Sc to Zn, there is a general reduction in atomic and ionic sizes as the increasing nuclear charge becomes more effective. Applied to hydration enthalpies, this factor implies that they would be expected to become more negative as the ionic size decreases across the series. Superimposed on this trend are electronic effects that may be understood by a consideration of the effects of an octahedral arrangement of six water molecules around a charged metal centre.

The variations in ionic radii of the transition elements of the 4th period serve to exemplify the arguments needed to rationalize similar variations in the other transition series. Figure 7.3 includes a plot of the radii of the $2+$ ions of those transition elements of the 4th period that form them. The plot includes the radius of the Ca^{2+} ion, which represents the beginning of the series but has no 3d electrons, as also has the Zn^{2+} and Ga^{3+} ions (both $3d^{10}$) at the end of the series. The radii are those of octahedrally coordinated ions as they are found in crystalline compounds, the counter-ions (*i.e.* the ions of opposite charge) being situated at the vertices of an octahedron, as shown in Figure 7.4.

There is a general downward trend in the radii going across the period, but the dips at V^{2+}, Ni^{2+}, V^{3+} and Co^{3+}, and the general shape of the plot, may be explained in terms of which d orbitals are occupied in each case.

The molecular orbital theory of the coordinate bond makes use of the five nd, the single $(n+1)$s and the three $(n+1)$p orbitals of the central ion of a complex ion, where n is the value of the principal quantum number, equal to 3 for the first series of transition elements. Assuming that the coordinate bonding between the central ion and the six water ligands is σ-type, only the $3d_{z^2}$ and $3d_{x^2-y^2}$ set of d orbitals (e_g) and the 4s (a_{1g}) and 4p (t_{1u}) orbitals are of the correct symmetry to overlap with the ligand group orbitals, which transform as $a_{1g} + e_g + t_{1u}$. The $3d_{xy}$, $3d_{xz}$ and $3d_{yz}$ set of d orbitals (t_{2g}) remain as non-bonding, since they do not have the correct symmetry to overlap effectively with any of the ligand group orbitals. These interactions are shown in the molecular orbital diagram of Figure 7.5.

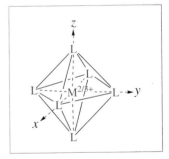

Figure 7.4 A cation surrounded octahedrally by six ligand atoms, L; in this text, L is taken to be the oxygen atom of a water molecule

The orbital envelopes of the 3d orbitals are shown in Figure 7.6. The $3d_{z^2}$ and $3d_{x^2-y^2}$ orbitals form the set labelled as e_g and the $3d_{xy}$, $3d_{xz}$ and $3d_{yz}$ orbitals form the set labelled as t_{2g}. Note that the e_g orbitals are directed along the coordinate axes and the t_{2g} orbitals bisect pairs of coordinate axes. The symmetry terminology is described in TCT no. 5, *Structure and Bonding*.[1]

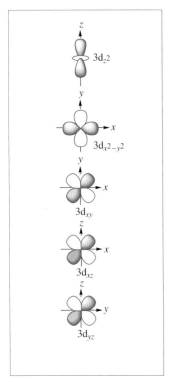

Figure 7.6   The envelopes of the 3d orbitals

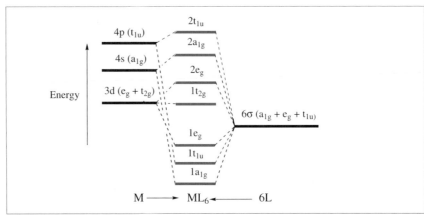

Figure 7.5   A molecular orbital diagram for octahedrally coordinated transition metal ions

The six electron pairs of the ligand water molecules occupy the six bonding orbitals, $1a_{1g} + 1e_g + 1t_{1u}$, and any remaining electrons must occupy the non-bonding $3d_{xy}$, $3d_{xz}$ and $3d_{yz}$ set ($1t_{2g}$), and possibly the anti-bonding orbitals that arise from the overlap of the $3d_{z^2}$ and $3d_{x^2-y^2}$ set (e_g) with the e_g ligand group orbitals and labelled as $2e_g$.

The energy gap between the $2e_g$ orbitals and the $1t_{2g}$ orbitals is given the value Δ_{oct} for octahedral complexes. A diagram showing the value of Δ_{oct} is given in Figure 7.7.

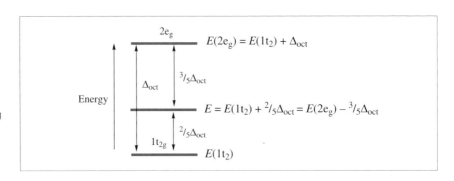

Figure 7.7   A diagram showing the relative energies of the $1t_{2g}$ and $2e_g$ molecular orbitals and the average energy for an electron in a spherical ligand field

The energies of any d configuration, d^n, may be referred to the particular average energy of the available orbitals when they are all singly occupied. Thus, the configuration $(1t_{2g})^3(2e_g)^2$ has an energy given by:

$$E[(1t_{2g})^3(2e_g)^2] = 3E(1t_{2g}) + 2E(2e_g)$$
$$= 3E(1t_{2g}) + 2(E(1t_{2g}) + \Delta_{oct})$$
$$= 5E(1t_{2g}) + 2\Delta_{oct} = 5(E(1t_{2g}) + {}^2/_5\Delta_{oct}) \quad (7.12)$$

So each electron could be considered to have the energy $E(1t_{2g}) + \frac{2}{5}\Delta_{oct}$. The position of this average energy, equivalent to the orbitals being surrounded by a spherical field rather than an octahedral one, is shown in Figure 7.7. The extra stabilization or destabilization that is associated with any particular d-electron configuration may be calculated from the occupancies of the $1t_{2g}$ and $2e_g$ orbitals. The stabilization energy is called **ligand field stabilization energy (LFSE)**, and the results for 0–10 d electrons are given in Table 7.8 and plotted in Figure 7.8.

Table 7.8 LFSE values for d-electron configurations in high-spin (HS) and low-spin (LS) cases; the units of energy are $\Delta_{oct}/5$

Numbers of electrons			LFSE
d	$1t_{2g}$	$2e_g$	
0	0	0	0
1	1	0	−2
2	2	0	−4
3	3	0	−6
4 (HS)	3	1	−3
4 (LS)	4	0	−8
5 (HS)	3	2	0
5 (LS)	5	0	−10
6 (HS)	4	2	−2
6 (LS)	6	0	−12
7 (HS)	5	2	−4
7 (LS)	6	1	−9
8	6	2	−6
9	6	3	−3
10	6	4	0

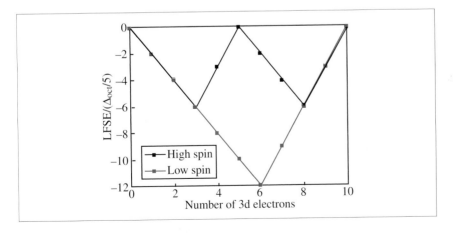

Figure 7.8 A plot of LFSE against d configuration for high-spin and low-spin cases

Hund's rules state that electrons entering degenerate orbitals occupy them singly, if possible, and have the maximum number of parallel spins.

The **Jahn–Teller effect** has a bearing on the ionic radii of Cr^{2+} in its high-spin state $(1t_2)^3(2e_g)^1$ and the Cu^{2+} ion $(1t_2)^6(2e_g)^3$. Both ions possess an odd filling of the $2e_g$ anti-bonding orbitals. The Jahn–Teller effect indicates that when an odd filling of a degenerate set of orbitals occurs, a distortion of the coordination arrangement will take place, with the effects of removing the degeneracy and stabilizing the structure. In the case of octahedral coordination, the doubly degenerate $2e_g$ levels have their degeneracy split by a tetragonal distortion along the z-axis, giving two long bonds, and four short bonds in the xy-plane. The $3d_{z^2}$ orbital is stabilized, becoming less anti-bonding. The $3d_{x^2-y^2}$ orbital is further destabilized, becoming more anti-bonding. In both cases under consideration, the $(1t_2)^3(2e_g)^1$ configuration of Cr^{2+} and the $(1t_2)^6(2e_g)^3$ configuration of Cu^{2+} are more stable as $(1t_2)^3 (3d_{z^2})^1$ and $(1t_2)^6(3d_{z^2})^2(3d_{x^2-y^2})^1$, respectively. [In this discussion the 3d orbitals of the $2e_g$ set indicate the metal's contribution to the anti-bonding orbitals.]

Because the $1t_{2g}$ orbitals are the next lowest in energy after the filled bonding orbitals, they are occupied by the first three available electrons. The three unpaired electrons occupy the triply degenerate level with parallel spins, as expected from **Hund's rules**. There is a choice for the fourth electron. It may pair up with one of the electrons in the $1t_{2g}$ level, or remain unpaired and enter the higher energy $2e_g$ set of orbitals. The factor that decides the fate of the fourth electron is the magnitude of Δ_{oct} compared to the energy of repulsion that it would suffer if it were to pair up in the $1t_{2g}$ level. A large value of Δ_{oct} favours electron pairing in the $1t_{2g}$ level; a small value allows the occupation of the upper $2e_g$ level. The value of Δ_{oct} depends largely on the nature of the ligand and the oxidation state of the central metal ion. Water is an intermediate ligand in its effect on the molecular orbital energies, and in all the aqua complexes of the $+2$ transition metal ions of the first series the maximum number of unpaired electrons in the $1t_{2g}$ and $2e_g$ levels occurs. For the $+3$ oxidation states of the first series, the aqua complexes are of maximum spin up to and including Fe^{3+}, but with Co^{3+} the value of Δ_{oct} is sufficiently greater than the pairing energy of repulsion to force electron pairing in the $1t_{2g}$ level so that the six d electrons all pair up in the $1t_{2g}$ level. Cases where the magnitude of Δ_{oct} is too small to force electrons to pair up in the t_{2g} level are called **high spin (HS)**. Where the opposite is the case, the configurations are called **low spin (LS)**.

The ligand field effects upon ionic radii can be understood in terms of the different shielding characteristics of the $2e_g$ and $1t_{2g}$ electrons. In the series of M^{2+} ions from Ca^{2+} to V^{2+}, electrons are progressively added to the $1t_{2g}$ orbitals, which lie along the bisectors of the coordinate axes. These electrons are not very effective at shielding the ligands from the effect of the increasing nuclear charge, and so the radii become progressively smaller. In the high-spin cases of Cr^{2+} and Mn^{2+}, the additional electrons are added to the $2e_g$ orbitals which, because they are mainly directed along the coordinate axes, are better at shielding the ligands from the nuclear charge than are electrons in the $1t_{2g}$ orbitals. The radius trend is reversed in the two ions, and then continues in the general downwards trend in the high-spin Fe^{2+}, $1t_{2g}^4 2e_g^2$, case where the added electron occupies the $1t_{2g}$ level. The high-spin Co^{2+} and Ni^{2+} ions show a decrease in radius as electrons are added to the $1t_{2g}$ orbitals. The remaining ions of the series (Cu^{2+} and Zn^{2+}) show an increasing radius trend as the additional electrons are added to the poorer shielding $2e_g$ orbitals. The ligand field effects on the radii of the M^{2+} ions explains the trends observed in their hydration enthalpies.

Table 7.9 contains the enthalpies of hydration of the M^{2+} ions from Ca^{2+} to Zn^{2+} with the exception of Sc^{2+}, which does not exist in aqueous solution, and Ti^{2+} for which evidence is dubious. The table

Table 7.9 Enthalpies of hydration and values of Δ_{oct} and LFSEs for the first transition series of $+2$ ions (enthalpies in kJ mol^{-1})

Ion	$\Delta_{hyd}H°$	$LFSE/(\Delta_{oct}/5)$	Δ_{oct}	LFSE	$\Delta_{hyd}H° - LFSE$
Ca^{2+}	-1616	0	0	0	-1616
V^{2+}	-1980	-6	151	-181	-1799
Cr^{2+}	-1945	-3	166	-100	-1845
Mn^{2+}	-1888	0	93	0	-1888
Fe^{2+}	-1988	-2	124	-50	-1938
Co^{2+}	-2051	-4	111	-89	-1962
Ni^{2+}	-2134	-6	102	-122	-2012
Cu^{2+}	-2135	-3	151	-91	-2044
Zn^{2+}	-2083	0	$-$	0	-2083

also contains the values of Δ_{oct}, derived from the electronic spectra of aqueous solutions containing the ions, and the corresponding values of LFSE.

Figure 7.9 shows a plot of the enthalpies of hydration for the Ca^{2+} to Zn^{2+} ions before and after adjustment for ligand field effects. The almost straight line with a negative slope indicates the effects of the variation in effective nuclear charge across the series. The plot of the actual values of the enthalpies of hydration against the 3d configurations of the ions may be visualized as the superposition of the high-spin plot of Figure 7.8 on the black line of Figure 7.9.

The same considerations apply to the M^{3+} ions from Sc^{3+} to Fe^{3+}. Table 7.10 contains the enthalpies of hydration of the M^{3+} ions from Sc^{3+} to Co^{3+} and that of Ga^{3+}. It also contains the values of Δ_{oct} (derived from the electronic spectra of aqueous solutions containing the ions) and the values of LFSE where appropriate.

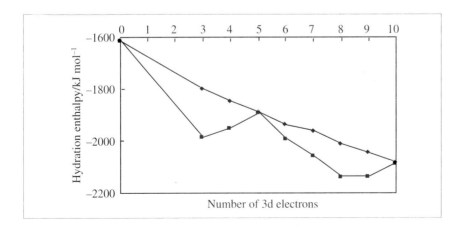

Figure 7.9 A plot of the enthalpies of hydration of some $+2$ ions against their 3d electronic configurations (*red line*) and their values adjusted for the effects of LFSE (*black line*)

Table 7.10 Enthalpies of hydration and values of Δ_{oct} and LFSEs for the first transition series of +3 ions (enthalpies in kJ mol^{-1})

Ion	$\Delta_{hyd}H°$	$LFSE/(\Delta_{oct}/5)$	Δ_{oct}	LFSE	$\Delta_{hyd}H° - LFSE$
Sc^{3+}	−3989	0	0	0	−3989
Ti^{3+}	−4296	−2	243	−97	−4199
V^{3+}	−4482	−4	212	−170	−4312
Cr^{3+}	−4624	−6	208	−250	−4373
Mn^{3+}	−4590	−3	251	−151	−4439
Fe^{3+}	−4486	0	164	0	−4486
Co^{3+}	−4713	−12	222	−533	−4180
Ga^{3+}	−4745	0	0	0	−4745

Figure 7.10 shows plots of their enthalpies of hydration against their 3d electronic configurations. The two graphs show similar variations superimposed on a general decreasing trend. The Sc^{3+} ion is a d^0 case and acts as the reference point for the discussion. A similar curve would be expected for the +3 ions from Co^{3+} to Ga^{3+}, but for the inconvenient experimental fact that no intermediate ions exist.

The bonding molecular orbitals are occupied by the six pairs of electrons from the six ligand water molecules, so the d electrons of the ions occupy the non-bonding 1t$_{2g}$ and anti-bonding 2e$_g$ orbitals as appropriate. The magnitude of Δ_{oct} is sufficiently small compared to the electron-pair repulsion energy to make the configurations of the ions from Ti^{3+} to Fe^{3+} high spin. For Co^{3+}, the six d electrons doubly occupy the three 1t$_{2g}$ non-bonding orbitals, because the gap between them and the anti-bonding 2e$_g$ orbitals is large enough to enforce electron pairing. The electronic occupancies of the 1t$_{2g}$ and 2e$_g$ orbitals are given in Table 7.11 for the ions under discussion.

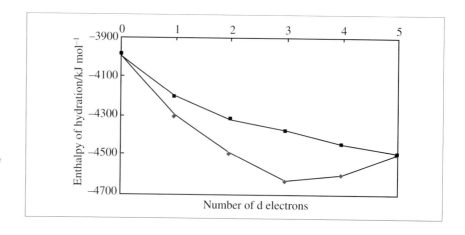

Figure 7.10 A plot of the enthalpies of hydration of some +3 ions (*red line*) against their 3d electron configurations and their values adjusted for the effects of LFSE (*black line*)

Across the series, electrons are progressively added to the non-bonding t_{2g} orbitals, and there is a consequent increase in the negative values of the enthalpies of hydration of the $+3$ ions as far as Cr^{3+} as the nuclear charges become more effective. In Mn^{3+}, one electron enters the $2e_g$ level, and the consequent weakening of the bonding offsets the increasing nuclear charge to make the enthalpy of hydration of the ion less negative than that of the Cr^{3+} ion. This trend is repeated with Fe^{3+}, with its $(2e_g)^2$ configuration. In the Co^{3+} ion, all six d electrons occupy the non-bonding $1t_{2g}$ orbitals, leading to stronger metal–water bonding and a considerable increase in the negative value of the enthalpy of hydration of the ion.

Table 7.11 d-electron configurations of the $+3$ ions from Sc to Co and Ga

Ion	Number of $1t_{2g}$ electrons	Number of $2e_g$ electrons
Sc^{3+}	0	0
Ti^{3+}	1	0
V^{3+}	2	0
Cr^{3+}	3	0
Mn^{3+}	3	1
Fe^{3+}	3	2
Co^{3+}	6	0
Ga^{3+}	6	4

7.4 Variations of Ionic Forms and Redox Behaviour across the First Series of Transition Elements

In this section the redox properties of the 10 elements of the first transition series are discussed. The lower oxidation states ($+2$ and $+3$, $+1$ of Cu) of these elements are treated together and the higher states are described and discussed separately.

7.4.1 The Lower Oxidation States of the Elements of the First Transition Series, Sc–Zn

Table 7.12 contains the Latimer-type data for the reduction potentials up to and including the $+3$ oxidation states of the first series of transition elements.

Table 7.12 Some reduction potentials for the first series of transition elements at pH $= 0$

			$E°$ /V at pH $= 0$			
III		II		I		0
Sc^{3+}			-2.03			Sc
Ti^{3+}			-1.21			Ti
V^{3+}	-0.255	V^{2+}		-1.13		V
Cr^{3+}	-0.424	Cr^{2+}		-0.9		Cr
Mn^{3+}	$+1.5$	Mn^{2+}		-1.18		Mn
Fe^{3+}	$+0.771$	Fe^{2+}		-0.44		Fe
Co^{3+}	$+1.92$	Co^{2+}		-0.282		Co
		Ni^{2+}		-0.257		Ni
		Cu^{2+}	$+0.159$	Cu^+	$+0.52$	Cu
		Zn^{2+}		-0.762		Zn

Table 7.13 Values of E° (M^{3+}/M) for the elements V to Co

Redox couple	E° (M^{3+}/M)/V
V^{3+}/V	−0.84
Cr^{3+}/Cr	−0.74
Mn^{3+}/Mn	−0.29
Fe^{3+}/Fe	−0.04
Co^{3+}/Co	+0.45

Worked Problem 7.5

Q   From the data given in Table 7.12, calculate the values of the standard reduction potentials for the V^{3+}/V couple.

A   The value of E° for the reduction of the M^{3+}(aq) ion to the solid metal, M, may be calculated from the values of E° for the reduction of the M^{3+}(aq) ion to M^{2+}(aq) and for the reduction of the M^{2+}(aq) ion to the metal, M. The value of $E^\circ(V^{3+}/V)$ is calculated from $E^\circ(V^{3+}/V^{2+})$ and $E^\circ(V^{2+}/V)$ by taking into account the numbers of electrons associated with the reduction processes:

$$E^\circ(V^{3+}/V) = [E^\circ(V^{3+}/V^{2+}) + 2 \times E^\circ(V^{2+}/V)] \div 3$$
$$= (-0.255 + 2 \times -1.13)/3$$
$$= -0.84 \text{ V (to two decimal places)}$$

By similar calculations, the values of $E^\circ(M^{3+}/M)$ for the elements Cr–Co are those given in Table 7.13, together with that for V.

7.4.2   Trends in the M^{2+}/M Reduction Potentials

Table 7.14 contains the observed and calculated values of E° for the reductions of the +2 ions to their respective metals. It also contains the thermochemical data related to the transformation of the solid metal atoms to M^{2+} ions in solution. The calculated values of the potentials are those based upon the standard enthalpies of formation of the +2 ions in aqueous solution. The entropy terms are neglected and this omission is justified from the close agreement between the calculated and observed values of the potentials given in Table 7.12. The enthalpies of hydration of the +2 ions are those given in Table 7.5.

Table 7.14   Data (all in kJ mol^{-1}) for Ca and the first-row transition elements, and the calculated standard reduction potentials for the reduction of M^{2+} to the element, M

Element (M)	Δ_aH° (M)	I_1	I_2	$\Delta_{hyd}H^\circ$ (M^{2+})	E° /V (calc)	E° /V (obs)
Ca	+178	590	1145	−1616	−2.81	−2.87
V	+514	651	1414	−1980	−1.25	−1.13
Cr	+397	653	1591	−1945	−0.75	−0.91
Mn	+281	717	1509	−1888	−1.15	−1.18
Fe	+416	762	1562	−1988	−0.46	−0.44
Co	+425	760	1648	−2051	−0.30	−0.28
Ni	+430	737	1753	−2134	−0.28	−0.26
Cu	+337	745	1958	−2135	+0.34	+0.34
Zn	+130	906	1733	−2083	−0.80	−0.76

The plots of the experimental and calculated values of $E°(M^{2+}/M)$ shown in Figure 7.11 indicate that they agree very well and that this means that the main factors influencing the values have been identified.

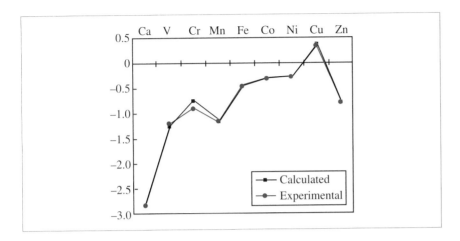

The $+2/0$ reduction potentials have enthalpic contributions from the terms in the equation:

$$E°(M^{2+}/M) = -[-\Delta_{hyd}H°(M^{2+}) - I_2 - I_1 - \Delta_a H°(M) + (2 \times 420)] \div 2F \qquad (7.13)$$

where the $+420\,kJ\,mol^{-1}$ term corresponds to the enthalpy of oxidation of half of one mole of dihydrogen to give one mole of hydrated protons, and the factor of two relates to the two-electron reduction process.

This treatment indicates that the magnitude of the reduction potential is governed by the values of two relatively large terms, $(-\Delta_{hyd}H°(M^{2+}) + 840)$ and $(I_1 + I_2)$, and the smaller term $\Delta_a H°(M)$. The interplay between the three terms decides the value for the $E°(M^{2+}/M)$ potential. Greater negative values of $\Delta_{hyd}H°(M^{2+})$, associated with smaller crystal radii of the $+2$ ions, contribute to a more negative value of $E°(M^{2+}/M)$. Greater sums of the first and second ionization energies have an opposing effect and contribute to a less negative value of $E°(M^{2+}/M)$. The less important $\Delta_a H°(M)$ term contributes in a manner similar to the ionization energy sum; the greater its value, the less negative the reduction potential becomes. The more negative the value of $E°(M^{2+}/M)$, the more difficult it is to reduce the $+2$ ion to the metal; the less negative the value, the easier it is to reduce the $+2$ ion to the metal. Figure 7.12 shows plots of the atomization energies of the first-row transition metals and that of calcium, the sums of their first two ionization energies, and the negative values of their $+2$ enthalpies of hydration.

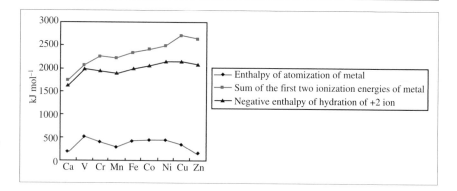

Figure 7.12 Plots of $\Delta_a H^\circ$ (M), $I_1 + I_2$, and $-\Delta_{hyd}H^\circ(M^{2+})$ for calcium and the first-row transition elements

The general increase in the ionization energy sum is dominant in deciding the trend towards less negative values of the $+2/0$ reduction potentials. Superimposed on the trend are the particularly low values of atomization enthalpy for Ca, Mn and Zn. The more positive values of the standard reduction potentials are favoured by: (i) a less negative value for the enthalpy of hydration of the $+2$ ion, (ii) a high value for the sum of the first two ionization energies of the metal, and (iii) a high value of the enthalpy of atomization of the metal.

Equation (7.13) may be rearranged to read as:

$$E^\circ(M^{2+}/M) = -[-\Delta_{hyd}H^\circ(M^{2+}) + 840] + [I_2 + I_1 + \Delta_a H^\circ(M)] \div 2F$$

(7.14)

The terms $[-\Delta_{hyd}H^\circ(M^{2+}) + 840]$ tend to make the value of $E^\circ(M^{2+}/M)$ negative and the terms $[I_1 + I_2 + \Delta_a H^\circ(M)]$ have the opposite effect. The values of the two composite terms are plotted in Figure 7.13 for the elements from Ca–Zn.

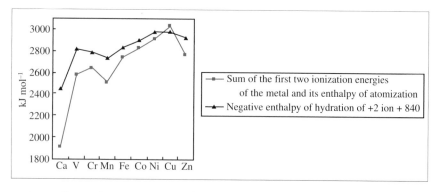

Figure 7.13 Plots of the sums of the first two ionization energies of the metal and its enthalpy of atomization and the negative enthalpy of the enthalpy of hydration of the $+2$ ion, to which has been added 840 kJ mol^{-1}

Apart from the copper case, the two terms interact to make the value of $E^\circ(M^{2+}/M)$ negative, the term $[-\Delta_{hyd}H^\circ(M^{2+}) + 840]$ having a value that is greater than that of the term $[I_1 + I_2 + \Delta_a H^\circ(M)]$. In the copper case the values of the two terms are reversed, making $E^\circ(Cu^{2+}/Cu)$ positive.

The unique behaviour of Cu in the first transition series, having a positive $E^\circ(M^{2+}/M)$ value, accounts for its inability to liberate H_2 from dilute acids (*i.e.* those with a molar concentration of 1 mol dm^{-3}). Only oxidizing acids (concentrated nitric and hot concentrated sulfuric) react with Cu, the acids being reduced. The reason for this behaviour is the high energy needed to transform Cu(s) to Cu^{2+}(g), which is not exceeded by the hydration enthalpy of the ion.

The data in Table 7.13 show that there is a considerable variation in the values of the enthalpies of atomization of the elements. These are a measure of the metallic bond energy, which depends upon the number of valence electrons and their distribution in the valence bands of molecular orbitals (as discussed in TCT no. 9, *Atomic Structure and Periodicity*[2]). Low values contribute to a more negative value for the corresponding reduction potential. The values of the enthalpies of atomization are, numerically, the least important in the calculation of the reduction potential. There is a general trend towards less negative values for the reduction potentials along the series, with a dip at Mn. This corresponds to the particularly low value of the enthalpy of atomization of Mn and its low second ionization energy. The less negative enthalpy of hydration offsets these effects somewhat, but not sufficiently to avoid the observed and calculated dips in the reduction potential values of Mn. The stability, due to maximization of the exchange energies between electrons of like-spin, of the half-filled d sub-shell in Mn^{2+} and the completely filled d^{10} configuration in Zn^{2+}, are related to their E° values, whereas E° for Ni is related to the highest negative $\Delta_{hyd}H^\circ$, corresponding to its smaller radius.

7.4.3 Reactions of the First-row Transition Elements with Dilute Acid

Across the sequence from V to Co there is a general trend in the values of the M^{3+}/M reduction potentials, given in Table 7.13, from the negative value of -0.84 V for V to the positive value of 0.45 V for Co. This has the implication that dissolving the metals in dilute acid will produce their +3 states, except for Co, since the value of $E^\circ(Co^{3+}/Co)$ is positive. Oxidation of Co by H^+ would stop at the +2 state. The production of the +3 states, although thermodynamically viable for the metals from V to Fe, does not necessarily occur. The mechanism of solution of metals in acidic solutions has not been studied to any great extent, but it is probable that the reactions occur in one-electron stages. The interaction of the aqueous protons and the metal surface may be considered to be a transfer of electrons to the protons to produce hydrogen atoms, which would then dimerize to give dihydrogen. The oxidized metal atoms could be released

Simultaneous transfers of more than one electron between energy levels have low probabilities. They may be induced spectroscopically by intense laser radiation of suitable photon energy, but are rare in chemical reactions.

into the bulk of the solution with charges consistent with the ion's stability in aqueous conditions, or possibly could be released in a singly ionized state to give a transient $M^+(aq)$ species:

$$M(s) + H^+(aq) \rightarrow M^+(aq) + H(aq) \qquad (7.15)$$

$$2H(aq) \rightarrow H_2(g) \qquad (7.16)$$

Of the transition elements, only silver has a water-stable singly charged cation, $Ag^+(aq)$. Copper does have a stable +1 ion in solid compounds, but this disproportionates in aqueous solution:

$$2Cu^+(aq) \rightarrow Cu(s) + Cu^{2+}(aq) \qquad (7.17)$$

to give the metal and the water-stable +2 state. Mercury has a water-stable +1 state, but this has the formula Hg_2^{2+} and consists of two mercury(I) atoms bonded together covalently.

The copper disproportionation reaction suggests that $M^+(aq)$ ions of the transition elements might undergo the same fate:

$$2M^+(aq) \rightarrow M(s) + M^{2+}(aq) \qquad (7.18)$$

The direct production of higher charged cations from a metal surface is a possibility, but it is more likely that the +1 ions are first released and subsequently disproportionate to give the water-stable +2 ions. Any further oxidation by aqueous protons would then occur, if thermodynamically feasible, in the bulk solution:

$$M^{2+}(aq) + H^+(aq) \rightarrow M^{3+}(aq) + \tfrac{1}{2}H_2(g) \qquad (7.19)$$

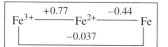

A Latimer diagram for the +2 and +3 states of iron is shown in the margin. Although the slightly negative value for the +3/0 potential implies that metallic iron should be oxidized to the +3 state by aqueous protons, such a reaction does not occur. Pure iron wire dissolves readily in dilute sulfuric acid in the *absence* of dioxygen to give a solution of $Fe^{2+}(aq)$. The reaction is slow at room temperature and, like most chemical reactions, goes faster with gentle heating. Only if dioxygen is present is there any further oxidation of the iron. This example shows the necessity of understanding mechanisms of reactions in addition to their thermodynamics in order to understand their chemistry. The initial product of the reaction between metallic iron and aqueous protons is probably Fe^+, which then disproportionates to give Fe and Fe^{2+}. The aqueous proton cannot bring about further oxidation to Fe^{3+}, and a fairly strong oxidizing agent is necessary, one with a reduction potential greater than +0.77 V.

The reactions of the Sc–Zn metals with dilute acid solutions, together with some comments about the preparation of the +2 and +3 states, are summarized in Table 7.15.

Table 7.15 Oxidation of the transition elements from Sc to Zn

Element	Means of oxidation
Sc	Tarnishes in air; reacts readily with dilute acid solutions to give $Sc^{3+}(aq)$
Ti	Metal protected by oxide film; oxidized by hot concentrated HCl to the (III) state
V	Metal protected by oxide film; oxidized by concentrated HNO_3 to (IV) and (V) oxides
Cr	Metal protected by oxide film; reacts very slowly with dilute acid solution to give $Cr^{3+}(aq)$
Mn	Reacts readily with dilute acid solution to give $Mn^{2+}(aq)$
Fe	Reacts readily with dilute acid solution to give $Fe^{2+}(aq)$
Co	Reacts slowly with dilute acid solution to give $Co^{2+}(aq)$
Ni	Reacts readily with dilute acid solution to give $Ni^{2+}(aq)$
Cu	No reaction with dilute acid; oxidized by nitric acid to $Cu^{2+}(aq)$
Zn	Tarnishes in air; reacts readily with dilute acid solutions to give $Zn^{2+}(aq)$

7.4.4 Variations in $E^\circ(3+/2+)$ for some 4th Period Transition Elements

Not all the first-row transition elements have the same pairs of stable oxidation states. The elements from V to Co are chosen for this example because they all have $+2$ and $+3$ states for which reliable experimental data are available. The data and the calculated E° values are shown in Table 7.16. The reduction enthalpies are calculated by using the equation:

$$\text{Reduction enthalpy} = -\Delta_{hyd}H^\circ(M^{3+}) - I_3(M) + \Delta_{hyd}H^\circ(M^{2+}) + 420$$

$$(7.20)$$

The calculated value for the reduction potential is obtained by dividing the reduction enthalpy by $-F$.

Table 7.16 Data (all in kJ mol^{-1}) for the elements V to Co of the calculated standard reduction potentials for the reduction of M^{III} to M^{II}

	V	Cr	Mn	Fe	Co
$\Delta_{hyd}H^\circ(M^{3+})$	−4482	−4624	−4590	−4486	−4713
$I_3(M)$	2828	2987	3248	2957	3232
$\Delta_{hyd}H^\circ(M^{2+})$	−1980	−1945	−1888	−1988	−2051
E° (calculated)	−0.97	−1.16	+1.31	+0.40	+1.56
E° (experimental)	−0.26	−0.42	+1.54	+0.77	+1.92

The trends in the terms $\Delta_{hyd}H^\circ(M^{3+})$, $I_3(M)$ and $\Delta_{hyd}H^\circ(M^{2+})$ are shown in Figure 7.14. They may be understood from a consideration of the changes in electronic configurations.

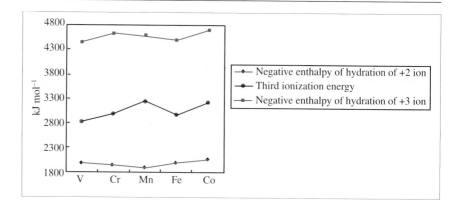

Figure 7.14 Plots of $\Delta_{hyd}H°(M^{3+})$, $I_3(M)$ *and* $\Delta_{hyd}H°(M^{2+})$ for the elements V, Cr, Mn, Fe and Co

Across the sequence of elements, V to Co, there is a general reduction in atomic and ionic sizes as the increasing nuclear charge becomes more effective. Applied to hydration enthalpies, this factor implies that they would be expected to become more negative as the ionic size decreases across the series. Superimposed on this trend are electronic effects that may be understood by a consideration of Figure 7.5. The water ligands supply all the bonding electrons, and the electrons of the metal ions are then accommodated in first the $3d_{xy}$, $3d_{xz}$ and $3d_{yz}$ ($1t_{2g}$) non-bonding orbitals, which are occupied singly in accordance with Hund's rules. Any extra electrons occupy the $3d_{z^2}$ and $3d_{x^2-y^2}$ anti-bonding orbitals ($2e_g$), because in aqua complexes of the $+2$ ions the difference in energy between the anti-bonding orbitals and the non-bonding orbitals is small. The occupation of the anti-bonding $2e_g$ orbitals affects some of the M^{2+} ions, *i.e.* Cr^{2+} $(2e_g)^1$ and Mn^{2+}, Fe^{2+} and Co^{2+} [all with $(2e_g)^2$] are destabilized to some extent, the anti-bonding electrons causing the metal–oxygen (of the water ligands) bonding to be weakened, with consequent increase in size.

The same consideration applies to the M^{3+} ions Mn^{3+} $(2e_g)^1$ and Fe^{3+} $(2e_g)^2$, which have less negative hydration enthalpies as a consequence. The higher stability (more negative hydration enthalpy) of the Co^{3+} ion arises because the six d^6 electrons occupy the three non-bonding $1t_{2g}$ orbitals, because the gap between them and the anti-bonding $2e_g$ orbitals is large enough to enforce electron-pairing. There is, in consequence, stronger metal–water bonding in the aquated Co^{3+} ion. The trends in the third ionization energies are understandable as a general increase across the series, but with the values for iron and cobalt being lower than expected because the electron removal occurs from a doubly occupied orbital, whereas those from vanadium, chromium and manganese occurs from singly occupied orbitals. Plots of the experimental and calculated values of $E°(M^{3+}/M^{2+})$ for the first-row transition elements are shown in Figure 7.15.

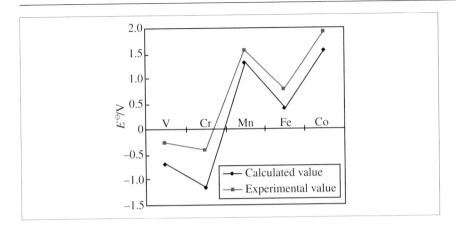

Figure 7.15 Plots of the experimental and calculated values of $E^{\circ}(M^{3+}/M^{2+})$ for some of the first-row transition elements

The trends in the calculated values are similar to those in the observed data. In the calculations, the omission of the entropy terms has caused some large differences between the observed and calculated values for the M^{3+}/M^{2+} potentials.

It is clear from the examples in this section and those of Section 7.4.2 that quite simple calculations can identify the major factors that govern the values of E° for any couple of monatomic states, and that major trends may be rationalized.

7.5   General Redox Chemistry of the d-Block Elements

Using a variety of Latimer and volt-equivalent diagrams, this section consists of a general survey of the redox chemistry of the elements of the d-block of the Periodic Table. The diagrams in the margins for each group of elements give their Group number at the top, their atomic numbers as a left-hand superscript to the element symbol, their Allred–Rochow electronegativity coefficients as a right-hand subscript, and their outer electronic configurations as free atoms beneath the element symbol.

Although there is no space to develop a detailed discussion of the solubilities of compounds of the transition elements, the general insolubility of their $+2$ and $+3$ hydroxides is important. The rationale underlying their insolubility can be summarized: (i) the hydroxide ion is relatively small (152 pm ionic radius) and the ions of the $+2$ and $+3$ transition metals assume a similar size if their radii are increased by 60–80 pm, and (ii) the enthalpy of hydration of the hydroxide ion (-519 kJ mol^{-1}) is sufficiently negative to represent a reasonable degree of competition with the metal ions for the available water molecules, thus preventing the metal ions from becoming fully hydrated. Such effects combine to allow the lattice enthalpies of the hydroxides to become dominant.

In transition metal chemistry, there are some oxides that are insoluble in acidic solutions (e.g. MnO_2) and considerably more that are insoluble in alkaline solutions (e.g. TiO_2, Mn_2O_3, MnO_2, VO, and V_2O_3). The oxides are the thermodynamically preferred form of the particular oxidation states in the solid form, most other metals forming hydroxides in alkaline conditions. Oxide formation is governed by the larger negative lattice enthalpies provided by the doubly charged oxide ion compared to the singly charged hydroxide ion.

Group 3	
44.956	
Sc	
21	1.2
$4s^2 3d^1$	
88.906	
Y	
39	1.1
$5s^2 4d^1$	

138.906	174.967
La	**Lu**
57 1.1	71 1.1
$6s^2 5d^1$	$6s^2 4f^{14} 5d^1$

+3		0
Sc^{3+}	−2.03	Sc
Y^{3+}	−2.37	Y
La^{3+}	−2.38	La
Lu^{3+}	−2.30	Lu

Group 4
47.867
Ti
22   1.3
$4s^2 3d^2$
91.224
Zr
40   1.2
$5s^2 4d^2$
178.49
Hf
72   1.2
$6s^2 4f^{14} 5d^2$

By contrast, the chlorides of the metal ions are soluble because the chloride ion (181 pm ionic radius) is considerably larger than the hydroxide ion, and its enthalpy of hydration (-359 kJ mol^{-1}) is less negative than that of OH$^-$. This allows the metal cations to exert more nearly their full effect on the solvent molecules, thus overcoming the lattice enthalpy terms, and this leads to their general solubility as chlorides.

The insolubility of the hydroxides of the lower oxidation states of the transition elements is the reason for the general lack of aqueous chemistry in alkaline solutions. The higher oxidation states of the elements take part in covalency to produce oxoanions and persist even in alkaline conditions, and allow their solubility.

7.5.1 Group 3 Redox Chemistry

Group 3 of the Periodic Table consists of the elements scandium, yttrium and either lanthanum or lutetium, depending upon the preferred arrangement of the Table. Group 3 elements have the outer electronic configuration $ns^2 np^1$, and invariably their solution chemistry is that of the $+3$ state. In this text, treatment of both La and Lu is carried out in Chapter 8, which deals with the f-block elements. Lanthanum and lutetium represent the first and last members of the lanthanide series.

The elements of the Group, including La and Lu, are powerful reducing agents, and their $+3/0$ standard reduction potentials in 1 mol dm^{-3} acid solution are summarized by the Latimer diagrams shown in the margin.

The elements have no basic solution chemistry; their $+3$ oxidation states have the form of hydroxides or oxides in alkaline conditions.

7.5.2 Group 4 Redox Chemistry

The redox chemistry of titanium, zirconium and hafnium in 1 mol dm^{-3} acid solution is summarized by the Latimer diagrams:

+4		+3		+2		0
TiO^{2+}	+0.1	Ti^{3+}	−0.37	Ti^{2+}	−1.63	Ti
Zr^{4+}			−1.55			Zr
Hf^{4+}			−1.7			Hf

The Ti(II) state is not well characterized in aqueous solution, but exists in the solid as the dichloride. If the Ti^{3+}/Ti^{2+} potential of -0.37 is correct, the $+2$ ion is on the edge of the region in which the reduced form should reduce water. The only states of Zr and Hf in aqueous solution are the $+4$ ions. Their reduction potentials imply that the metals are good reducing agents. The $+4$ ions are extensively hydrolysed in weak acid solutions and tend to form polymeric ions. There is no basic solution chemistry of note.

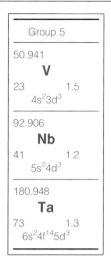

7.5.3 Group 5 Redox Chemistry

Volt-equivalent diagrams for the oxidation states of V are given in Figure 7.16 for pH values of 0 and 14. The reduction potentials on which the diagrams are based are given in the margin as a vertical Latimer diagram.

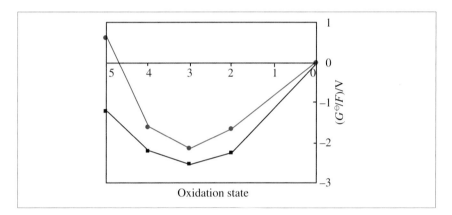

Figure 7.16 Volt-equivalent diagrams for the oxidation states of vanadium at pH $= 0$ (*red line*) and pH $= 14$ (*black line*)

   The most stable state of the vanadium is V^{3+} in acidic solution, the $+4$ and $+5$ states existing as oxocations, VO^{2+} and VO_2^{+} respectively. The $+5$ state has a quite large reduction potential with respect to the most stable $+3$ state, indicating its oxidizing power. If a solution of VO_2^{+} is mixed with one containing V^{2+}, the $+5$ state oxidizes the $+2$ state to the most stable state in acid solution, $+3$, and is itself reduced to that state:

$$VO_2^{+}(aq) + 2V^{2+}(aq) + 4H^{+}(aq) \rightarrow 3V^{3+}(aq) + 2H_2O(l) \qquad (7.21)$$

This type of reaction, the opposite of *disproportionation*, is called **comproportionation**. In such reactions, two oxidation states of the same element undergo reduction and oxidation to produce an intermediate state with greater thermodynamic stability than either of the reactants.

   The maximum stability for vanadium as the V^{3+} ion in acidic solution can be understood in terms of the maximization of the enthalpy of hydration for the $+3$ ion, above which hydrolysis alters the form and stability of the higher states. The $+4$ and $+5$ states are considerably electronegative compared to the lower oxidation states, and are able to engage in covalent bonding to ligand oxide ions to form V=O bonds.

pH $= 0$		pH $= 14$
VO_2^{+}		VO_4^{3-}
$+1.00$		$+2.19$
VO^{2+}		$HV_2O_5^{-}$
$+0.34$		$+0.54$
V^{3+}		V_2O_3
-0.255		-0.486
V^{2+}		VO
-1.13		-0.82
V		V

There is some loss of overall stability, which gives the two higher states their oxidizing properties.

In alkaline solution the $+5$ state has the form VO_4^{3-}, which has greater oxidant properties than at pH $= 0$. It is easily reduced to the most stable V^{III} oxide, the driving force being the large lattice energy of the solid oxide. The high lattice energy of the solid oxide is a better contribution to the exothermicity of the reduction process in alkaline solution than is the enthalpy of hydration energy of the V^{3+} ion in acidic solution. The $+4$ state has the formula $HV_2O_5^-$, a dimeric hydrolysis product of the acidic version, VO^{2+}. The pH $= 14$ volt-equivalent plot shown in Figure 7.16 has the same form as the plot at pH $= 0$. The lower oxidation states are more stabilized by their lattice energies being greater than the hydration enthalpies of the ions that are soluble in acid solution. The higher states undergo hydrolysis and make use of covalency, but their stabilities decrease as the oxidation state increases. At all pH values the highest oxidation state is that corresponding to the participation of all five valency electrons of the metal.

The Latimer diagrams of Nb and Ta in 1 mol dm^{-3} acid solution are:

$+5$		$+3$		0
Nb_2O_5	-0.1	Nb^{3+}	-1.1	Nb
Ta_2O_5		-0.81		Ta

They represent the thermodynamic data for the stability of the $+5$ oxides with respect to their formation from their elements. The Nb^{3+} ion is not well characterized. In alkaline solutions, both Nb and Ta form polymeric anions of the formulae $[H_xM_6O_{19}]^{(8-x)-}$ with values for x of 0, 1, 2 or 3.

7.5.4 Group 6 Redox Chemistry

A volt-equivalent diagram for the oxidation states of Cr is shown in Figure 5.4 for a pH value of 0. It shows the same general form as the diagram for vanadium in Figure 7.16, except that the highest oxidation state is $+6$. The more stable states are the cations with low oxidation states, and as the oxidation state increases the ionic forms become less and less stable as oxocations. The most stable state of chromium is the $+3$ state, Cr^{3+}, in acidic solution. The highest oxidation state, $+6$, is a powerful oxidizing agent. The $+2$ state has the potential to reduce water to dihydrogen, but the reaction is very slow and solutions of Cr^{2+} may be prepared by the reduction of the $+3$ state with zinc amalgam, which are

Group 6	
51.996	
Cr	
24	1.6
$4s^1 3d^5$	
95.94	
Mo	
42	1.3
$5s^1 4d^5$	
183.84	
W	
74	1.4
$6s^2 4f^{14} 5d^4$	

reasonably stable in the absence of dioxygen. The $+4$ and $+5$ states are unstable and undergo disproportionation to the more stable $+3$ and $+6$ states.

A vertical Latimer diagram for $pH = 14$ is shown in the margin. In alkaline solution the $+6$ state loses its oxidizing properties. Chromium(VI) in alkaline solution appears as the monomeric CrO_4^{2-} ion, which is much smaller than the $Cr_2O_7^{2-}$ ion, but without a change in the overall charge. The enthalpy of hydration of the Cr(VI) species produced is more than doubled, and this is the main determining factor causing the loss of oxidizing capacity of the $+6$ state in alkaline solution. The lattice energy of the insoluble Cr(III) hydroxide is not so important.

The aqueous chemistry of molybdenum and tungsten is complicated by polymer formation in acid solution and reduction potential data are not known with certainty. The acid-solution chemistry of molybdenum is summarized in Table 7.17.

pH = 14	
$+6$	CrO_4^{2-}
	-0.11
$+3$	$Cr(OH)_3$
	-1.33
0	Cr

Table 7.17 The oxidation states of molybdenum in acid solution

Oxidation state	Simple ions	Bridged ions
$+2$	–	$[(H_2O)_4Mo_2(H_2O)_4]^4$
$+3$	$[Mo(H_2O)_6]^{3+}$	$[(H_2O)_4Mo(\mu\text{-}OH)_2Mo(H_2O)_4]^{4+}$
$+4$	–	$[Mo_3(\mu_3\text{-}O)(\mu\text{-}O)_3(H_2O)_9]^{4+}$
$+5$	–	$[Mo_2O_2(\mu\text{-}O)_2(H_2O)_6]^{2+}$
$+6$	–	$[MoO_2(OH)(H_2O)_3]^{2+}$

The $[(H_2O)_4Mo_2(H_2O)_4]^{4+}$ ion possesses a quadruple bond between the two $+2$ state atoms; its structure is shown in Figure 7.17.

Molybdenum in its $+2$ state has a $4d^4$ electronic configuration, with the eight electrons of the atoms engaging in quadruple bond formation with σ-type overlaps of the two d_{z^2} orbitals along the Mo–Mo axis, and the two sets of d_{xz} and d_{yz} orbitals overlapping in the xz and yz planes, respectively, in a π manner. That arrangement leaves the two sets of d_{xy} and $d_{x^2-y^2}$ orbitals that are perpendicular to the Mo–Mo axis, and which transform as δ orbitals in the linear ion-molecule. One of the sets is used to make the fourth Mo–Mo bond and the other set, together with the vacant 5s, $5p_x$ and $5p_y$ orbitals of both atoms, are used to accept electron pairs from the two sets of four water molecules that contribute to the hydrated ion by coordinate bonding.

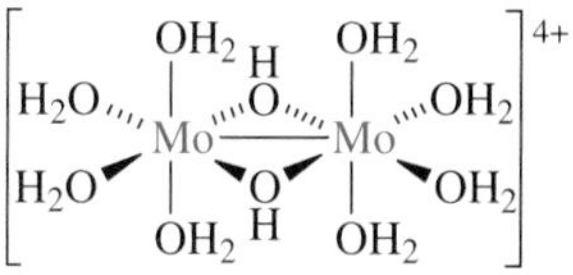

The structure of the double hydroxo-bridged $[(H_2O)_4Mo(\mu\text{-}OH)_2Mo(H_2O)_4]^{4+}$ ion is shown in Figure 7.18. There are two OH bridging

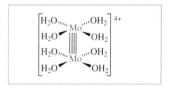

Figure 7.17 The structure of the $[(H_2O)_4Mo_2(H_2O)_4]^{4+}$ ion; the Mo – Mo bond has a bond order of four: a quadruple bond

Figure 7.18 The structure of the bridged $[(H_2O)_4Mo(\mu\text{-}OH)_2Mo (H_2O)_4]^{4+}$ ion

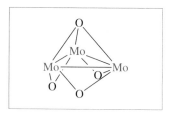

Figure 7.19 The basic structure of the $[Mo_3(\mu_3\text{-}O)(\mu\text{-}O)_3(H_2O)_9]^{4+}$ ion; there are also three water molecules coordinated to each of the Mo atoms

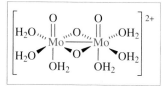

Figure 7.20 The structure of the $[Mo_2O_2(\mu\text{-}O)_2(H_2O)_6]^{2+}$ ion

Polyoxometallates are polyoxoanions based upon the polymerization of simple oxoanions of metallic elements. They are formed by the higher oxidation states of the elements of Groups 4, 5, 6 and 7, with particular proliferation in the chemistry of V, Mo and W.

groups between the two Mo^{III} atoms (signified by the μ symbol) and a direct Mo–Mo single bond, making use of the two sets of the available d^3 electrons.

The +4 state ion $[Mo_3(\mu_3\text{-}O)(\mu\text{-}O)_3(H_2O)_9]^{4+}$ has the structure shown in Figure 7.19, which is based on an Mo_3 triangle with single Mo–Mo bonds. In this formula the μ_3-O symbolism indicates that three molybdenum atoms share the apical oxygen atom. The μ-O indicates that the oxygen atom bridges two Mo atoms. Three oxygen atoms bridge the three pairs of adjacent Mo atoms, and the fourth oxygen atom caps the Mo_3 triangle on the side opposite to the three Mo–O–Mo bridges and is bonded to all three metal atoms. There are three water molecules attached to each of the Mo atoms to complete the structure.

The +5 state ion $[Mo_2O_2(\mu\text{-}O)_2(H_2O)_6]^{2+}$ has the structure shown in Figure 7.20 and contains a Mo–Mo single bond with two bridging oxygen atoms.

There are no well-characterized simple aqueous ions of tungsten in acid solution, but the ions $W_3O_4^{4+}$ and $W_2O_4^{2+}$ exist and probably have structures similar to those given for the Mo ions in Figures 7.19 and 7.20, respectively.

In alkaline solution, molybdenum and tungsten in their +6 states form MO_4^{2-} ions and these are the most stable states, as indicated by the Latimer-type table that includes the insoluble +4 oxides:

+6		+4		0
MoO_4^{2-}	−0.78	MoO_2	−0.98	Mo
WO_4^{2-}	−1.26	WO_2	−0.98	W

In dilute acid solutions, molybdenum and tungsten form many polyoxometallate ions.

A general equation for the production of polyoxometallate ions is:

$$x MO_4^{n-}(aq) + y H^+(aq) \rightleftharpoons [M_xO_{4x-y/2}]^{(xn-y)-}(aq) + y/2 H_2O(l) \quad (7.22)$$

Examples are $[Mo_7O_{24}]^{6-}$ and $[W_{10}O_{32}]^{4-}$. There are also many examples where the polyoxometallate ions contain a number of protons, e.g. $[H_3Mo_7O_{24}]^{3-}$ and $[H_2W_{12}O_{42}]^{10-}$.

7.5.5 Group 7 Redox Chemistry

The redox chemistry of manganese is dealt with volt-equivalent diagrams and a description of the small amount of aqueous chemistry of Tc and Re follows. A volt-equivalent diagram for the oxidation states of Mn is shown in Figure 7.21 for pH values of 0 and 14. The values of the potentials are given in a vertical Latimer diagram in the margin.

The most stable state of manganese is Mn^{2+} in acidic solution, and the highest oxidation state, MnO_4^-, is an oxidizing agent; it is easily reduced to the +2 state. Unlike the V and Cr cases, the most stable +2 state of Mn occurs because of the maximization of exchange energy in its d^5 configuration.

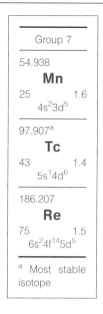

Group 7
54.938
Mn
25   1.6
$4s^2 3d^5$
97.907^a
Tc
43   1.4
$5s^1 4d^6$
186.207
Re
75   1.5
$6s^2 4f^{14} 5d^5$
^{a} Most stable isotope

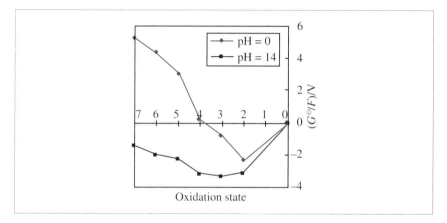

Figure 7.21 Volt-equivalent diagrams for the oxidation states of manganese at pH = 0 and 14

The +3 ion is unstable in acid solution with respect to disproportionation to give the +7 and +2 states. The reaction seems not to include the +4 state as an intermediate, since that state is not water soluble (MnO_2). The +6 and +5 states are unstable in aqueous acid solutions, and disproportionate into the soluble +7 state and solid MnO_2.

In alkaline solution the only important state is the +7 oxoanion, MnO_4^-, which has the same form as in acid solution and retains its powerful oxidizing properties. There are different reduction products at the two pH values. In acidic solution, MnO_4^- is a powerful oxidant and is reduced to Mn^{2+} without going through the intermediate stage of producing the insoluble MnO_2. In alkaline solution the end product of the reduction of MnO_4^- is the insoluble MnO_2. The +5 ion MnO_4^{3-} disproportionates, but the +6 state MnO_4^{2-} is more stable in alkaline than it is in acid solution.

Technetium is a synthetic element, unknown in Nature. Its most useful isotope is produced by neutron bombardment of the ^{98}Mo stable isotope in the form of the MoO_4^{2-} ion to give the TcO_4^- ion via a "neutron-in, beta minus particle-out" process [$^{98}Mo(n,\beta^-)^{99m}Tc$].

The technetium isotope produced is the metastable ^{99m}Tc, which is in a nuclear excited state. The eventual fall to the ground state has a half-life of 6 hours, and is accompanied by the emission of a gamma-ray photon. The gamma-ray photons have energies sufficiently low not to harm

pH = 0	pH = 14
MnO_4^-	MnO_4^-
+0.90	+0.56
$HMnO_4^-$	MnO_4^{2-}
+1.28	+0.27
H_3MnO_4	MnO_4^{3-}
+2.9	+0.93
MnO_2	MnO_2
+0.95	+0.15
Mn^{3+}	Mn_2O_3
+1.54	−0.23
Mn^{2+}	$Mn(OH)^2$
−1.19	−1.56
Mn	Mn

An electronic configuration is stabilized by the exchange energy between electrons of the same spin. The d^5 high-spin arrangement in Mn^{2+}(aq) is stabilized since all five electrons have the same spin.

Stable nuclei have neutron/proton ratios (N/P) within a particular range of values. Beta-minus decay occurs when nuclei are unstable because their neutron/proton ratio is too large. A neutron is transformed into a proton plus an electron, and the electron is ejected from the nucleus as a β^- particle. The original N/P ratio decreases to $(N-1)/(P+1)$ and the daughter nucleus is either stable or less unstable than the parent.

Group 8		
55.847		
Fe		
26		1.6
$4s^2 3d^5$		
101.07		
Ru		
44		1.5
$5s^1 4d^7$		
190.2		
Os		
76		1.5
$6s^2 4f^{14} 5d^6$		

human tissue, but high enough to allow the isotope to be used medicinally for diagnostic purposes when injected into the body. The ^{99}Tc ground state is a soft β^- emitter so does not threaten the well being of the patients.

The aqueous chemistry of Re is that of the $+7$ state ion ReO_4^-, which does not have the oxidizing power possessed by the Mn^{VII} equivalent ion.

7.5.6 Group 8 Redox Chemistry

The Latimer diagrams for iron in 1 mol dm^{-3} H$^+$ solution and 1 mol dm^{-3} OH$^-$ solution are:

pH	$+6$		$+3$		$+2$		0
1			Fe^{3+}	$+0.77$	Fe^{2+}	-0.44	Fe
14	FeO_4^{2-}	$+0.81$	Fe_2O_3	-0.86	$Fe(OH)_2$	-0.89	Fe

The chemistry of iron in aqueous solution is dominated by the $+2$ and $+3$ states, which are well characterized. The $+3$ state in acid solution is a good oxidizing agent; the $+2$ state is the most stable. The $[Fe(H_2O)_6]^{3+}$ complex ion is a violet colour in the solid chlorate(VII) salt, but in solution it undergoes hydrolysis to give the familiar orange-red colour. The first stage of the hydrolysis may be written as:

$$[Fe(H_2O)_6]^{3+} \rightleftharpoons [Fe(H_2O)_5OH]^{2+} + H^+ \qquad (7.23)$$

for which the pK value at 25 °C is 2.74. In the pH range 1–1.8, more hydrolysis occurs and products include the μ-oxo dimer, $[(H_2O)_5FeOFe(H_2O)_5]^{4+}$, produced by the elimination of water between two $[Fe(H_2O)_5OH]^{2+}$ ions. At pH values above 1.8 the hydrated $+3$ oxide is precipitated. Transient species containing Fe^{IV} and Fe^V have been reported, but there is no aqueous chemistry of note.

In alkaline conditions the $+2$ and $+3$ states are found in solids, the $+3$ existing as the hydrated oxide, $Fe_2O_3.xH_2O$. If $x = 1$, the formula simplifies to FeO(OH) and if $x = 3$ it becomes $Fe(OH)_3$. The substance dissolves in concentrated NaOH solution to give the FeO_2^- ion. Fe^{III} in concentrated KOH solution may be oxidized to the $+6$ state using OCl^-. The blue salt K_2FeO_4 can be prepared in such a manner; it contains the FeO_4^{2-} ion and in alkaline solution is a moderate oxidizing agent.

A volt-equivalent diagram for the oxidation states of Ru is shown in Figure 7.22 for acid solution. The values of the potentials used to construct the diagram are given in the vertical Latimer diagram in the margin.

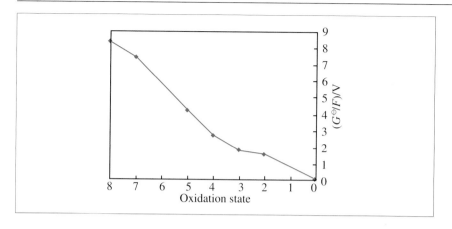

Figure 7.22 A volt-equivalent diagram for ruthenium at pH = 0

Thermodynamically, the elementary state is the most stable oxidation state of ruthenium; the higher oxidation states are all at higher Gibbs energy values. Previous examples of volt-equivalent diagrams for V, Cr and Mn, for example, have shown a common trend with the lower oxidation states being more stable than the elements. Only when the oxidation state increases beyond the $+2/+3$ region does the stability of the higher oxidation states decrease. Ruthenium is sometimes classified together with osmium from Group 8, rhodium and iridium from Group 9, and palladium and platinum from Group 10 as the **platinum metals.** They exist together naturally as metals or as sulfide minerals. All six metals are extremely resistant towards oxidation, and severe conditions must be used to cause their solution. Ruthenium can be obtained in aqueous solution after treatment with a fused mixture of sodium hydroxide and either sodium peroxide or potassium chlorate(V).

A Latimer diagram for the $+4$ and $+8$ states of Os in acid solution is:

$+8$		$+4$		0
OsO_4	$+1.02$	OsO_2	$+0.65$	Os

The $+8$ oxide is a molecular compound, unlike many transition metal oxides that have giant lattice arrangements, and is quite soluble in water. It has considerable oxidizing properties and is used as an oxidizing agent in many organic reactions. The $+4$ oxide is insoluble in water.

7.5.7 Group 9 Redox Chemistry

The Group 9 elements, cobalt, rhodium and iridium, have redox chemistry which in aqueous acidic solution can be summarized by Latimer diagrams:

Oxidation state	pH = 0
$+8$	RuO_4
	$+0.99$
$+7$	RuO_4^-
	$+1.6$
$+5$	RuO_2^+
	$+1.5$
$+4$	$H_n[Ru_4O_6^- (H_2O)_{12}]^{(4+n)+}$
	$+0.86$
$+3$	Ru^{3+}
	$+0.24$
$+2$	Ru^{2+}
	$+0.8$
0	Ru

The "nobility" of the platinum metals, compared to the relative reactivity of the members of their groups in the 4th period, Fe, Co and Ni, is caused by their enthalpies of atomization, which are very large. These, together with the sums of the first three ionization energies, dominate the potential values, which are all positive and indicate that the metals will not dissolve in dilute acid. Their enthalpies of hydration are insufficiently negative to make the M^{3+}/M reduction potentials negative. The standard enthalpies of atomization of Fe, Co, Ni and the platinum metals are shown below, with units of kJ mol^{-1}.

	Fe	Co	Ni
$\Delta_a H°$	+418	+427	+431
	Ru	**Rh**	**Pd**
$\Delta_a H°$	+640	+556	+390
	Os	**Ir**	**Pt**
$\Delta_a H°$	+782	+665	+565

+3		+2		0
Co^{3+}	+1.92	Co^{2+}	−0.28	Co
Rh^{3+}		+0.76		Rh
Ir^{3+}		+1.0		Ir

Group 9	Group 10
58.933	58.69
Co	**Ni**
27 1.7	28 1.8
$4s^2 3d^7$	$4s^2 3d^8$
102.906	106.42
Rh	**Pd**
45 1.5	46 1.4
$5s^1 4d^8$	$4d^{10}$
192.22	195.08
Ir	**Pt**
77 1.6	78 1.4
$6s^2 4f^{14} 5d^7$	$6s^1 4f^{14} 5d^9$

Only the +2 state of cobalt has thermodynamic stability in acid solution. The instability of Co^{3+} is referred to in Section 5.3. Only the +3 states of Rh and Ir are stable in acid solution; their +3/0 standard reduction potentials are quite positive, consistent with their "nobility". In alkaline solutions the +2 and +3 states of the elements exist as insoluble hydroxides.

7.5.8 Group 10 Redox Chemistry

Of the Group 10 elements, nickel, palladium and platinum, only the +2 states of Ni and Pd are well characterized in aqueous acid solutions. Their +2/0 standard reduction potentials in acid solution are given in the Latimer diagram:

+2		0
Ni^{2+}	−0.257	Ni
Pd^{2+}	+0.915	Pd

In alkaline conditions, the +2 states are found only in the solid compounds $Ni(OH)_2$, PdO and PtO.

7.5.9 Group 11 Redox Chemistry

Of the Group 11 elements, copper, silver and gold, only Cu and Ag have ions that are well characterized in acid solutions. Copper and silver form

+3	+2		+1		0
	Cu^{2+}	+0.159	Cu^+	+0.52	Cu
			Ag^+	+0.799	Ag
Au_2O_3		+1.2	Au^+	+1.69	Au

+1 ions and copper alone has a water-stable +2 state. The Latimer diagrams summarize their standard reduction potentials:

The Cu^{2+}/Cu reduction potential (+0.34 V) represents the positive electrode of the Daniell cell (see Sections 4.1 and 7.5.10). All three members of the group are insoluble in dilute acid solutions. In alkaline conditions the solid compounds $Cu(OH)_2$, Cu_2O and Ag_2O represent the oxidation states of the elements.

The +1 Oxidation States of Cu, Ag and Au

The metals of Group 11 all form +1 states that vary in their stability with respect to the metallic state. The standard reduction potentials for the couples Cu^+/Cu and Ag^+/Ag are +0.52 V and +0.8 V, respectively. That for Au^+/Au has an estimated value of +1.62 V. The thermodynamic data for the calculation of the reduction potentials are given in Table 7.18, which also contains the calculated potentials for Cu and Ag.

	Group 11
63.546	**Cu**
29	1.8
	$4s^13d^{10}$
107.868	**Ag**
47	1.4
	$5s^14d^{10}$
196.967	**Au**
79	1.4
	$6s^14f^{14}5d^{10}$

Table 7.18 Data for Cu and Ag and the calculated reduction potentials for their M^+/M couples

	Δ_aH° /kJ mol^{-1}	I_1/kJ mol^{-1}	$\Delta_{hyd}H^\circ$/kJ mol^{-1}	E°(calc)/V
Cu	+337	745	−590	+0.75
Ag	+289	732	−490	+1.10
Au	+366	890	?	—

The reduction enthalpies are calculated by using the equation:

$$\text{Reduction enthalpy} = -\Delta_{hyd}H^\circ(M^+, g) - I_1(M) - \Delta_aH^\circ(M, g) + 420$$

$$(7.24)$$

Dividing the reduction enthalpy by F and then changing the sign of the result gives the calculated reduction potential.

The higher enthalpy of atomization and higher first ionization energy contribute to the high value of the reduction potential of the gold couple. The calculated values are not very different from the observed values, indicating that ignoring the entropy terms is not a large source of error. The **nobility** of gold, i.e. its high resistance to acids, is represented by the large positive reduction potential of the +1 state. The gold potential indicates that the +1 oxidation state would oxidize water and therefore would not be stable in aqueous solution. It is also unstable with respect to disproportionation into the metal and the +3 oxide.

Gold can be attacked by *aqua regia* [three volumes of concentrated HCl to one volume of concentrated nitric(V) acid] or by acidic solutions containing complexing agents, e.g. thiourea, $S=C(NH_2)_2$. *Aqua regia* produces the gold as the Au^{III} complex ion $[AuCl_4]^-$, and thiourea solutions produce the soluble Au^I complex ion $[(thiourea)_2Au]^+$. It is the combination of oxidant plus complexing agent that overcomes the nobility of the metal. With *aqua regia* the oxidant is nitric(V) acid, the complexing agent is the chloride ion. With thiourea, the organic compound is the complexing agent, and the oxidant is atmospheric dioxygen.

Worked Problem 7.6

Q Compare the reduction potentials of the K^+/K and Cu^+/Cu couples using observed and calculated values. Identify the factors responsible for the differences in the calculated values.

A The respective reduction enthalpies for the half-reactions:
$$K^+(aq) + e^- \rightarrow K(s) \text{ and } Cu^+(aq) + e^- \rightarrow Cu(s)$$
are calculated using equation (7.24). For K, the values of the quantities in the equation are: $340 - 419 - 89 + 420 = 252$ kJ mol^{-1}.

The value of the reduction enthalpy converts (by dividing by $-F$) to an E° of -2.61 V. For Cu, the values of the quantities in the equation are: $590 - 745 - 337 + 420 = -72$ kJ mol^{-1}.

The value of the reduction enthalpy converts to an E° of $+0.75$ V. The large difference between the two values can be seen to be due to the very much larger enthalpy of atomization of Cu, the much higher value of its first ionization energy, and that these two amounts are not compensated by the more exothermic enthalpy of hydration of the Cu^+ ion. The two couples have respective changes in electronic configuration:
$$K^+ [Ar] \rightarrow K [Ar]4s^1 \text{ and } Cu^+ [Ar]3d^{10} \rightarrow Cu [Ar]3d^{10}4s^1$$
and the changes in ionic/metallic radii are 138/230 pm for the K couple and 77/128 pm for the Cu couple. The smaller ionic radius for the Cu^+ ion, compared to that for K^+, arises from the poor shielding offered by the $3d^{10}$ electrons. The consequent effect upon the radii does not outweigh the greater values of the enthalpy of atomization and first ionization energy of Cu compared to the corresponding values for K.

7.5.10 Group 12 Redox Chemistry

Of the Group 12 elements, zinc, cadmium and mercury, only Hg has a water-stable $+1$ state, and all three elements have $+2$ states that are water-stable. Their reduction potentials are summarized in the Latimer diagram:

Group 12
65.38 **Zn** 30   1.7 $4s^23d^{10}$
112.41 **Cd** 48   1.5 $5s^24d^{10}$
200.59 **Hg** 80   1.5 $6s^24f^{14}5d^{10}$

+2		+1		0
Zn^{2+}			-0.762	Zn
Cd^{2+}			-0.402	Cd
Hg^{2+}	$+0.91$	Hg_2^{2+}	$+0.796$	Hg

The Zn^{2+}/Zn couple forms the negative electrode of the Daniell cell (see Section 4.1). The enthalpy changes that accompany the various stages of the reaction occurring in the Daniell cell are shown in Figure 7.23. The two aqueous ions are shown at the same level, as would be expected if there were no junction potentials connected to the operation of the salt bridge. The overall enthalpy change of -219 kJ mol^{-1} equates to a standard reduction potential of $(-219/-F) = +1.13$ V. The difference of 0.03 V between the calculated enthalpy-only value and the experimental E° value of 1.1 V is because of the small overall entropy change of -20.9 J K^{-1} mol^{-1}, which equates to a $T\Delta S^\circ$ value of -6.2 kJ mol^{-1} and to a contribution of -0.03 V to the value of E° for the overall process.

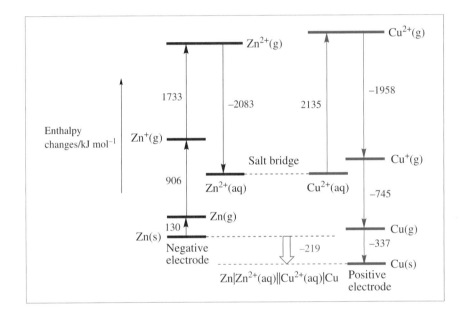

Figure 7.23 Enthalpy changes for the stages in the Daniell cell reaction

Zinc and cadmium dissolve in dilute acid to give their $+2$ ions, but mercury does not dissolve, as indicated by the two positive reduction potentials. Mercury forms the diatomic Hg_2^{2+} ion, in which the Hg–Hg bond length is 251 pm, consistent with it being a single σ bond formed from the overlap of the two 6s atomic orbitals. The reason for the relatively greater stability of the 6s electrons of Hg is relativistic stabilization which causes the first two ionization energies (1010 and 1810 kJ mol^{-1}) to be considerably greater than those of Zn (908 and 1730 kJ mol^{-1}) and Cd (866 and 1630 kJ mol^{-1}).

In alkaline solution, only Zn forms an ion, ZnO_2^{2-}, with a reduction potential to the metal of -1.22 V. In dilute alkaline solutions, $Zn(OH)_2$ precipitates, but it re-dissolves in stronger alkaline solution to form the

Although they do not exist in aqueous solution, the $+1$ ions Zn_2^{2+} and Cd_2^{2+} can be produced in the solid state: zinc metal dissolves in molten $ZnCl_2$ to give a glass with the formula Zn_2Cl_2, and cadmium dissolves in molten $Cd(AlCl_4)_2$ to give crystalline $Cd_2(AlCl_4)_2$. The Cd–Cd distance is 258 pm, consistent with a low bond strength.

zincate ion, formulated as ZnO_2^{2-} or $[Zn(OH)_4]^-$. Cadmium forms a hydroxide, $Cd(OH)_2$, and mercury(II) precipitates as the oxide, HgO.

Summary of Key Points

1. The ionic forms of transition elements in aqueous solution were described and the variation from cations to oxoanions was discussed.

2. The enthalpies of hydration of a selection of transition metal ions were derived.

3. The variations of values of the standard reduction potentials for M^{2+}/M, M^{3+}/M and M^{3+}/M^{2+} couples were explained in terms of enthalpies of atomization, ionization energies and enthalpies of hydration.

4. The redox chemistry of the transition elements was described group by group.

References

1. J. Barrett, *Structure and Bonding*, RSC Tutorial Chemistry Text, no. 5, Royal Society of Chemistry, Cambridge, 2001.
2. J. Barrett, *Atomic Structure and Periodicity*, RSC Tutorial Chemistry Text, no. 9, Society of Chemistry, Cambridge, 2002.
3. U. Müller, *Inorganic Structural Chemistry*, Wiley, New York, 1992. All ionic/metallic/covalent radii have been taken from this book to ensure consistency.

Further Reading

The first two books mentioned above in the References.
D. M. P. Mingos, *Essential Trends in Inorganic Chemistry*, Oxford University Press, Oxford, 1998.
F. A. Cotton, G. Wilkinson, C. A. Murillo and M. Bochmann, *Advanced Inorganic Chemistry*, 6th edn., Wiley, New York, 1999.

Problems

7.1. The standard enthalpy of formation of the $Cr^{2+}(aq)$ cation is -143.5 kJ mol^{-1}, the standard enthalpy of formation of gaseous chromium atoms is $+397$ kJ mol^{-1} and the first two ionization energies of chromium are 653 and 1591 kJ mol^{-1}. Calculate a value for the absolute standard enthalpy of hydration of the Cr^{2+} ion, using data for the proton reduction half-reaction given in Chapter 2.

7.2. The standard enthalpy of formation of the $Cu^+(aq)$ cation is $+71.7$ kJ mol^{-1}, the standard enthalpy of formation of gaseous copper atoms is $+337$ kJ mol^{-1} and the first ionization energy of copper is 745 kJ mol^{-1}. Calculate a value for the absolute standard enthalpy of hydration of the Cu^+ ion.

7.3. The standard enthalpy of formation of the $Sc^{3+}(aq)$ cation is -614.2 kJ mol^{-1}, the standard enthalpy of formation of gaseous scandium atoms is $+378$ kJ mol^{-1} and the first three ionization energies of scandium are 633, 1235 and 2389 kJ mol^{-1}. Calculate a value for the absolute standard enthalpy of hydration of the Sc^{3+} ion.

7.4. The question refers to the couples Co^{3+}/Co^{2+}, Ag^+/Ag, Ni^{2+}/Ni and Zn^{2+}/Zn. (i) Which is the strongest oxidizing agent? (ii) Which is the weakest oxidizing agent? (iii) Which is the weakest reducing agent? (iv) Which is the strongest reducing agent? (v) Which ions are capable of being reduced by Ni?

7.5. Calculate the overall potential for the reaction:
$$Fe(s) + Ni^{2+}(aq) \rightleftharpoons Fe^{2+}(aq) + Ni(s)$$
Indicate the feasibility of the reaction and draw a diagram of the enthalpy changes occurring, assuming zero junction potentials, in a cell in which the reaction may take place. See Figure 7.23 and the data of Table 7.14. From the enthalpy calculation, give a value for E° for the reaction.

7.6. Which oxidation states of the elements Ti–Co in the first transition series are most stable in the presence of dioxygen at pH values of 0 and 14?

8

Periodicity of Aqueous Chemistry III: f-Block Chemistry

The aqueous chemistry of the two rows of f-block elements, the lanthanides (lanthanum to lutetium) and the actinides (actinium to lawrencium), are sufficiently different from each other to be dealt with in separate sections. Similarities between the two sets of elements are described in the actinide section.

Aims

By the end of this chapter you should understand:

- The redox chemistry of the lanthanide elements
- The redox chemistry of the actinide elements
- The similarities and differences between the two series of elements

---

8.1 The Lanthanides

Promethium does not occur naturally. It was first isolated from the fission products of uranium. The longest-lived isotope is ^{145}Pm (17.7 years half-life). A 1s electron falls into the nucleus and causes a proton to become a neutron; excess energy is released as a γ-ray photon. The decay process is called **electron capture**.

The lanthanide elements are the 15 elements from lanthanum to lutetium. Both La and Lu have been included to allow for the different versions of the Periodic Table, some of which position La in Group 3 as the first member of the third transition series and others that place Lu in that position. If Lu is considered to be the first element in the third transition series, all members of that series possess a filled shell $4f^{14}$ configuration. The outer electronic configurations of the lanthanide elements are given in Table 8.1.

Table 8.1 The outer electronic configurations of the lanthanide elements; they all possess a $6s^2$ pair of electrons

Element	Symbol	5d	4f	Element	Symbol	5d	4f
Lanthanum	La	1	0	Gadolinium	Gd	1	7
Cerium	Ce	0	2	Terbium	Tb	0	9
Praseodymium	Pr	0	3	Dysprosium	Dy	0	10
Neodymium	Nd	0	4	Holmium	Ho	0	11
Promethium	Pm	0	5	Erbium	Er	0	12
Samarium	Sm	0	6	Thulium	Tm	0	13
Europium	Eu	0	7	Ytterbium	Yb	0	14
				Lutetium	Lu	1	14

The oxidation states of the lanthanide elements are given in Table 8.2.

Table 8.2 Oxidation states of the lanthanide elements; those in red are the most stable states in aqueous solution

	+4	+4						+4						
+3	+3	+3	+3	+3	+3	+3	+3	+3	+3	+3	+3	+3	+3	+3
					+2	+2							+2	
La	Ce	Pr	Nd	Pm	Sm	Eu	Gd	Tb	Dy	Ho	Er	Tm	Yb	Lu

The lanthanides all have their $+3$ states as the stable species in acidic solutions, as indicated by the data given in Table 8.3. The $+3$ states are produced by the removal of the $6s^2$ pair of electrons plus either the single 5d electron or one of the 4f electrons. In this respect they behave like the members of Group 3, any additional ionization being normally unsustainable by either lattice production or ion hydration.

The abbreviations Ln = lanthanide and An = actinide are used for general purposes in this text.

Table 8.3 Reduction potential data for the lanthanide elements

E° / V at pH = 0

+4		+3		+2		0
		La^{3+}			-2.38	La
Ce^{4+}	$+1.72$	Ce^{3+}			-2.34	Ce
		Pr^{3+}			-2.35	Pr
		Nd^{3+}			-2.32	Nd
		Pm^{3+}			-2.29	Pm
		Sm^{3+}	-1.4	Sm^{2+}	-2.65	Sm
		Eu^{3+}	-0.34	Eu^{2+}	-2.86	Eu

(continued)

Table 8.3 *continued*

E° /V at pH = 0

Gd^{3+}		-2.28	Gd
Tb^{3+}		-2.31	Tb
Dy^{3+}		-2.29	Dy
Ho^{3+}		-2.33	Ho
Er^{3+}		-2.32	Er
Tm^{3+}		-2.32	Tm
Yb^{3+}	-1.04	Yb^{2+} -2.81	Yb
Lu^{3+}		-2.3	Lu

The very negative Ln^{3+}/Ln potentials are consistent with the electropositive nature of the lanthanide elements; their Allred–Rochow electronegativity coefficients are all 1.1 except for europium, which has a value of 1.0. The lighter elements of Group 3, Sc and Y, both have electronegativity coefficients of 1.3. The nearest p-block element to the lanthanides in these properties is magnesium; $E^{\circ}(Mg^{2+}/Mg) = -2.37$ V, and its electronegativity coefficient is 1.2.

Only cerium has a higher state that is stable in solution, Ce^{4+}, corresponding to the removal of all four valence electrons. The +4 state is a powerful oxidant, with a reduction potential to the +3 state that is very dependent upon the acid in which it is dissolved (*e.g.* in sulfuric acid the reduction potential is +1.44 V, in nitric(V) acid +1.61 V, and in chloric(VII) acid +1.70 V, indicating some ion pair formation in the first two examples). Praseodymium and terbium have +4 oxides, but these dissolve in acidic solution to give the corresponding +3 states and dioxygen.

Of the three lanthanides that form characterized +2 states, only Eu^{2+} is reasonably stable in acidic solution. The +2 states of Sm and Yb have +3/+2 potentials, which imply that they should reduce water to dihydrogen.

Worked Problem 8.1

Q The M^{3+}/M standard reduction potentials for Sm, Eu and Yb are not included in Table 8.2. Calculate their values from the data given in Table 8.2.

A For Sm, the Sm^{3+}/Sm potential is: $[-1.4 - (2 \times 2.65)] \div 3 = -2.23$ V

For Eu, the Eu^{3+}/Eu potential is: $[-10.34 - 1(2 \times 2.86)] \div 3 = -12.02$ V

For Yb, the Yb^{3+}/Yb potential is: $[-1.04 - (2 \times 2.81)] \div 3 = -2.22$ V

Going along the lanthanide elements in order of their atomic numbers, the values of the $+3/0$ reduction potentials show a very weak trend towards lower negative values, with a slight discontinuity at Eu. As with other elements treated, it may be shown that the values of the reduction potentials are governed by the interactions of the enthalpies of atomization of the elements and their first three successive ionization energies compensated by the enthalpies of hydration of their $+3$ ions. These interactions and the associated entropy changes produce the observed trend. The discontinuity at Eu is due to the advantage of attaining the $4f^7$ configuration in the element. The three-electron reduction of the next lanthanide ion, Gd^{3+}, to its zero oxidation state is achieved by the reappearance of the $5d^1$ configuration halfway along the series, and this does not carry the same exchange energy advantage as the change in the reduction of Er^{3+}.

The enthalpies of hydration of the lanthanides are given in Table 8.4 and show a regular increasing negative value with decreasing ionic radius.

Table 8.4   Data for the lanthanide elements and their enthalpies of hydration

	r_i	$\Delta_f H°(Ln^{3+}, aq)$	$\Delta_a H°(Ln, g)$	I_1	I_2	I_3	$\Delta_{hyd} H°(Ln^{3+}, g)$
La	103.2	− 709.4	+431.0	538	1067	1850	− 3335
Ce	101	− 700.4	+423.0	534	1047	1949	− 3393
Pr	99	− 706.2	+355.6	527	1018	2086	− 3433
Nd	98.3	− 696.6	+327.6	533	1035	2132	− 3464
Pm	97	− 660	+348.0	536	1052	2152	− 3488
Sm	95.8	− 691.1	+206.7	545	1068	2258	− 3509
Eu	94.7	− 605.6	+175.3	547	1085	2404	− 3557
Gd	93.8	− 687	+397.5	593	1167	1990	− 3575
Tb	92.3	− 698	+388.7	566	1112	2114	− 3619
Dy	91.2	− 696.5	+290.4	573	1126	2200	− 3626
Ho	90.1	− 707	+300.8	581	1139	2204	− 3672
Er	89	− 705	+317.1	589	1151	2194	− 3696
Tm	88	− 705.2	+232.2	597	1163	2285	− 3722
Yb	86.8	− 674.5	+152.3	603	1175	2417	− 3762
Lu	86.1	− 702.6	+427.6	524	1341	2022	− 3758

The values of $\Delta_{hyd} H°(Ln^{3+}, g)$ are plotted against the ionic radii of the Ln^{3+} ions in Figure 8.1; the ionic radii are those in decreasing order from La^{3+} to Lu^{3+}, the decrease being known as the **lanthanide contraction.** The contraction occurs as the result of 4f electrons not offering efficient shielding of the increasing nuclear charge, and because of relativistic effects that cause some contraction with increasing nuclear charge.

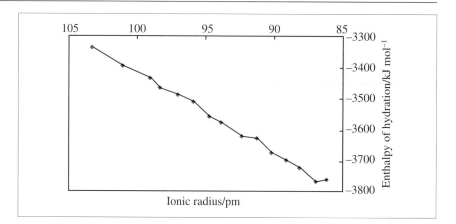

Figure 8.1 A plot of the enthalpy of hydration of Ln^{3+} ions against their ionic radii

---

Worked Problem 8.2

Q Calculate a value for the absolute standard enthalpy of hydration of the La^{3+} ion using the data from Table 8.3.

A The estimate of $\Delta_{hyd}H^{\circ}(La^{3+}, g)$ is given by the equation:

$$\begin{aligned}
\Delta_{hyd}H^{\circ}(La^{3+}, \text{ g}) = {} & \Delta_f H^{\circ}(La^{3+}, \text{ aq}) - \Delta_a H^{\circ}(La, \text{ g}) \\
& + 3\Delta_{hyd}H^{\circ}(H^+, \text{ g}) + 3I_H - I_1(La) \\
& - I_2(La) - I_3(La) + 3\Delta_a H^{\circ}(H, \text{ g}) \quad (8.1)
\end{aligned}$$

$$\begin{aligned}
= {} & -709.4 - 431 - (3 \times 1110) + (3 \times 1312) \\
& - 538 - 1067 - 1850 + (3 \times 218)
\end{aligned}$$

$$= -3335 \text{ kJ mol}^{-1}$$

Equation (7.14), adapted for the Ln^{3+}/Ln reduction potentials is:

$$\begin{aligned}
E^{\circ}(Ln^{3+}/Ln) = {} & -[-\Delta_{hyd}H^{\circ}(Ln^{3+}, \text{ g}) - I_3(Ln) - I_2(Ln) \\
& - I_1(Ln) - \Delta_a H^{\circ}(Ln, \text{ g}) + (3 \times 420)] \div 3F \quad (8.2)
\end{aligned}$$

It allows an estimate to be made of the standard reduction potentials of the lanthanide elements.

Worked Problem 8.3

Q Calculate the standard reduction potential for the La^{3+}/La couple from the data in Table 8.4.

A Using equation (8.2):

$$E^\circ(La^{3+}/La) = -[-\Delta_{hyd}H^\circ(La^{3+}, \text{ g}) - I_3(La) - I_2(La) - I_1(La)$$
$$- \Delta_a H^\circ(La, \text{ g}) + (3 \times 420)] \div 3F$$

$$= -[3335 - 1850 - 1067 - 538 - 431 + 1260] \div 3F$$

$$= -2.45 \text{ V}$$

The calculated and experimental values of the $+3/0$ standard reduction potentials for the lanthanide elements are given in Table 8.5.

Table 8.5 Calculated and experimental values of $E^\circ(Ln^{3+}/Ln)$

Ln	$E^\circ(Ln^{3+}/Ln)$ (calc)/V	$E^\circ(Ln^{3+}/Ln)$ (expt)/V	Ln	$E^\circ(Ln^{3+}/Ln)$ (calc)/V	$E^\circ(Ln^{3+}/Ln)$ (expt)/V
La	− 2.45	− 2.38	Tb	− 2.41	− 2.31
Ce	− 2.42	− 2.34	Dy	− 2.41	− 2.29
Pr	− 2.44	− 2.35	Ho	− 2.44	− 2.33
Nd	− 2.41	− 2.32	Er	− 2.44	− 2.32
Pm	− 2.28	− 2.29	Tm	− 2.43	− 2.32
Sm	− 2.39	− 2.23	Yb	− 2.33	− 2.22
Eu	− 2.09	− 2.02	Lu	− 2.43	− 2.30
Gd	− 2.38	− 2.22			

The calculated and experimental values for the standard reduction potentials agree very well, and the data may be used to isolate the factors that produce the very negative values. These are that the sums of the first three ionization energies and the enthalpies of atomization are generally low and are outbalanced by the considerably negative values of the enthalpies of hydration of the Ln^{3+} ions.

Worked Problem 8.4

Q The standard reduction potentials for the La^{3+}/La and Tl^{3+}/Tl couples are -2.38 V and $+0.74$ V, respectively. Identify the factors that are responsible for the large difference in the two values.

A The calculated value for $E^\circ(La^{3+}/La)$ is outlined in Worked Problem 8.3. Data for the two elements are given below (enthalpies in kJ mol^{-1}).

	r_i/pm	$\Delta_a H^\circ$(M, g)	I_1	I_2	I_3	$\Delta_{hyd}H^\circ$(M^{3+}, g)
La	104.5	+431	538	1067	1850	−3335
Tl	89	+182	589	1971	2878	−4164

Using equation (8.2) for thallium:

$$E^\circ(Tl^{3+}/Tl) = -[-\Delta_{hyd}H^\circ(Tl^{3+}, g) - I_3(Tl) - I_2(Tl) - I_1(Tl)$$
$$- \Delta_a H^\circ(Tl, g) + (3 \times 420)] \div 3F$$
$$= -[4164 - 2878 - 1971 - 589 - 182 + 1260] \div 3F$$
$$= +0.68V$$

The two calculated values are in good agreement with the experimental values, and the data must contain the explanation for the difference between the two results. A negative value for $E^\circ(Tl^{3+}/Tl)$ is favoured by the low enthalpy of atomization and the very negative value of the enthalpy of hydration term, but both the factors are over-ridden by the very much higher value of the sum of the ionization energies: 3455 kJ mol^{-1} for La, 5438 kJ mol^{-1} for Tl. The smaller ionic radius for Tl^{3+} produces the more negative enthalpy of hydration. The two ions have similar outer electronic configurations: La^{3+} has the [Xe] structure and Tl^{3+} is [Xe]$4f^{14}$. Thallium experiences the inert-pair effect, itself a consequence of the large values for the second and third ionization energies.

8.2   The Actinides

The actinide elements consist of the 15 elements from actinium to lawrencium. Their outer electronic configurations are given in Table 8.6.

Table 8.6   The outer electronic configurations of the actinide elements; the atoms all possess a 7s^2 pair of electrons

Element	Symbol	6d	5f	Element	Symbol	6d	5f
Actinium	Ac	1	0	Curium	Cm	1	7
Thorium[a]	Th	2	0	Berkelium	Bk	0	9
Protactinium	Pa	1	2	Californium	Cf	0	10
Uranium	U	1	3	Einsteinium	Es	0	11

(continued)

Table 8.6 *continued*

Element	Symbol	6d	5f	Element	Symbol	6d	5f
Neptunium	Np	1	4	Fermium	Fm	0	12
Plutonium	Pu	0	6	Mendelevium	Md	0	13
Americium	Am	0	7	Nobelium	No	0	14
				Lawrencium	Lr	1	14

[a]$6d^1 5f^1$ according to some sources.

As is the case for La and Lu, there is an argument for placing Lr as the first member of the fourth transition series, as it does possess a filled $5f^{14}$ set of orbitals.

The oxidation states of the actinide elements are given in Table 8.7.

Table 8.7 Oxidation states of the actinide elements; those in red are the most stable states in aqueous solution in the absence of dioxygen

				+7										
			+6	+6	+6	+6								
		+5	+5	+5	+5	+5								
	+4	+4	+4	+4	+4	+4	+4	+4	+4	+4	+4			
+3	+3	+3	+3	+3	+3	+3	+3	+3	+3	+3	+3	+3	+3	+3
+2	+2	+2	+2	+2	+2	+2	+2	+2	+2	+2	+2	+2		
Ac	Th	Pa	U	Np	Pu	Am	Cm	Bk	Cf	Es	Fm	Md	No	Lr

The +2, +3 and +4 ions are aquated ions with those charges and undergo hydrolysis to some extent in other than very acidic solutions. The +5 states all have the same ionic form as oxocations, AnO_2^+. In a similar manner the +6 states all have the ionic form as oxocations, AnO_2^{2+}.

Reduction potential data for the actinide elements, including lawrencium, are given in Table 8.8 for the well-characterized oxidation states.

Table 8.8 Reduction potential data for the actinides

$E°/V$ at $pH = 0$

+6		+5		+4		+3		+2	0
						Ac^{3+}		−2.13	Ac
				Th^{4+}			−1.83		Th
		PaO_2^+	−0.05	Pa^{4+}	−1.4	Pa^{3+}		−1.49	Pa
UO_2^{2+}	+0.17	UO_2^+	+0.38	U^{4+}	−0.52	U^{3+}		−1.66	U
NpO_2^{2+}	+1.24	NpO_2^+	+0.64	Np^{4+}	+0.15	Np^{3+}		−1.79	Np
PuO_2^{2+}	+1.02	PuO_2^+	+1.04	Pu^{4+}	+1.01	Pu^{3+}		−2.0	Pu

(continued)

Table 8.8 *continued*

$E° / V$ at $pH = 0$

AmO_2^{2+} $+1.6$ AmO_2^+ $+0.82$ Am^{4+} $+2.62$ Am^{3+}	-2.07			Am
Cm^{3+}	-3.7	Cm^{2+}	-1.2	Cm
Bk^{3+}	-2.8	Bk^{2+}	-1.54	Bk
Cf^{3+}	-1.6	Cf^{2+}	-1.97	Cf
Es^{3+}	-1.55	Es^{2+}	-2.2	Es
Fm^{3+}	-1.15	Fm^{2+}	-2.5	Fm
Md^{3+}	-0.15	Md^{2+}	-2.5	Md
No^{3+}	$+1.45$	No^{2+}	-2.6	No
Lr^{3+}	-2.1			Lr

Although some $+2$ states of the actinide elements exist, they have little stability in aqueous solution and are omitted from Table 8.8.

There are significant differences in the stable oxidation states in aqueous acidic solution compared with the corresponding elements of the lanthanide series. The underlying reason is that the 5f electrons of the actinide elements are more easily ionized than the 4f electrons of the corresponding lanthanides. In particular, there are differences in the first half of the series where the actinides form more stable oxidation states than do the lanthanides. At around the halfway stage, beyond curium, there is much more lanthanide-type behaviour of the later actinides. In general, the $+3$ state predominates, but it is not the most stable state in some cases. The elements that have the $+3$ state as their most stable state are Ac, Np, Pu, Am, Cm, Bk, Cf, Es, Fm, Md and Lr. Only four of the elements have other states as their most stable: Th, Pa, U and No. Thorium has a stable $+4$ state, consistent with the loss of all its valence electrons. Protactinium(V) is the most stable state for that element, again corresponding to the loss of all five valence electrons. U^{IV} and No^{II} are the most stable states of U and No. The $+5$ and $+6$ states of Pa, U, Np, Pu and Am exist in solution as the linear oxocations, AnO_2^+ and AnO_2^{2+}.

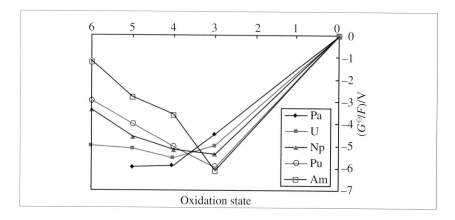

Figure 8.2 Volt-equivalent diagrams for the oxidation states of Pa, U, Np, Pu and Am at pH = 0

The potentials of Pa, U, Np, Pu and Am are displayed as volt-equivalent diagrams in Figure 8.2.

The most stable state of uranium can be seen to be the $+4$ state from the diagram of Figure 8.2. In uranium and subsequent elements, the most stable species is determined by the balance between the required ionization energy and the form of the ion after hydrolysis has occurred. Hydrolysis includes the stabilizing effects of the formation of metal–oxygen bonds as the electronegativity of the metal increases, encouraging covalent bond formation. In Th and Pa, the ionization energies to produce the maximum oxidations states are opposed by the stabilizations that occur with hydrolysis. In U, the balance changes somewhat, so the $+4$ state is the most stable. In the cases of Pu and Am the higher oxidation states become less and less stable, and in the subsequent (trans-americium) elements there is a general reversion to the $+3$ state as the most stable, as is the case with the lanthanide elements. With nobelium the most stable state is the $+2$ ion, which corresponds to the removal of the $7s^2$ pair, leaving the No^{2+} ion with the filled shell $5f^{14}$ configuration. Lawrencium behaves like lutetium, and is placed under that element in Group 3.

The actinide elements are all radioactive. Their nuclei are unstable mainly because of their very high charges; the attractive forces operating between the neutrons and protons do not fully balance the interproton repulsion. The nuclei consequently emit alpha particles, $^4He^{2+}$, to gain stability, *i.e.* to lose mass. Examples of radioactive decay are given in Table 8.9.

In α-particle decay the mass number changes from M to $M-4$ and the atomic number changes from A to $A-2$, the daughter element having two too many electrons. These undergo transfers to surrounding molecules so that eventually the helium nuclei (α-particles) become neutral helium atoms.

There are four radioactive decay series that, by sequences of α and β emissions, result in the eventual formation of a stable isotope of lead or bismuth. Members of each series have mass numbers which are exactly divisible by 4 (the thorium series, $4n$), or after being divided by four leave remainders of 1 (the neptunium series, $4n+1$), 2 (the uranium series, $4n+2$) or 3 (the actinium series, $4n+3$). The stable end-products of the four series are ^{206}Pb, ^{209}Bi, ^{207}Pb and ^{208}Pb, respectively.

Table 8.9   Radioactive properties of some actinide elements

An	Natural abundance (%)	$t_{1/2}$/y	α energy/MeV [a]
^{232}Th	100	1.4×10^{10}	4.08
^{235}U	0.72	7.04×10^{8}	4.68
^{238}U	99.27	4.47×10^{9}	4.04
^{239}Pu	—	2.41×10^{4}	5.24

[a]MeV = million electronvolts; 1 MeV = 1 GJ mol^{-1}.

The lanthanide elements are very difficult to separate because of their highly similar chemistry, but the earlier actinide elements have sufficiently different redox chemistry to allow easy chemical separations. This is important in the nuclear power industry, where separations have to be made of the elements produced in fuel rods of nuclear power stations as fission products, and of the products Np and Pu, which arise from the neutron bombardment of the uranium fuel.

Summary of Key Points

1. The redox chemistry of the lanthanide elements was described.

2. The enthalpies of hydration of the Ln^{3+} ions were calculated and their relationship to the lanthanide contraction was discussed.

3. The reduction potentials for Ln^{3+}/Ln couples were estimated and the factors identified which make them very negative.

4. The redox chemistry of the actinide elements was described.

5. The different ranges of oxidation states of the actinides in aqueous solution were described, discussed and compared with the much narrower ranges displayed by the lanthanide elements.

Further Reading

1. S. A. Cotton, *Lanthanides and Actinides*, Macmillan, London, 1991.
2. F. A. Cotton, G. Wilkinson, C. A. Murillo and M. Bochmann, *Advanced Inorganic Chemistry*, 6th edn., Wiley, New York, 1999.

Problems

8.1. From the data in Table 8.4, calculate a value for the absolute standard enthalpy of hydration of the Ho^{3+} ion.

8.2. Calculate an enthalpy-only value for the standard reduction potential for the Lu^{3+}/Lu couple from the data in Table 8.3.

8.3. From data given in the text, compare the standard reduction potentials of the $+1$, $+2$ and $+3$ oxidation states of the elements Cs, Ba and La, respectively, and their zero oxidation states. Identify the factors that are responsible for the observed trend in the values of the standard potentials.

Answers to Problems

Chapter 1

1.1. (i) The ratio m_S/m_W is given by $m_S\%/(100 - m_S\%)$. Let $m_S/m_W = f$. There are f kg of solute per kg of water. The volume of a solution consisting of f kg of solute and 1 kg of water, total mass $= (f + 1)$ kg, is given by: $(f + 1)/\rho$ m^3 or $1000(f + 1)/\rho$ dm^3. 1 dm^3 of the solution contains $f\rho/[1000(f + 1)]$ kg or $f\rho/(f + 1)$ g of the solute. The molar concentration is given by: $f\rho/[(f + 1) \times M_r(\text{solute})]$ mol dm^{-3}, where $M_r(\text{solute})$ is the relative molar mass of the solute.

(ii) The molality of the solution is given by $1000f/M_r(\text{solute})$.

(iii) The mole fraction of the solute is given by:

$$x_{\text{solute}} = \frac{f/M_r(\text{solute})}{[(f/M_r(\text{solute})) + (1/M_r(\text{water}))]}$$

1.2. The results are given in tabular form and derived by using the formulae of Problem 1.1:

Salt	Molar concentration/mol dm^{-3}	Molality, m	Mole fraction, x
Na$_2$SO$_4$	0.602	0.612	0.011
MgSO$_4$	0.917	0.923	0.0164

1 dm^3 of the Na$_2$SO$_4$ solution has a mass of 1071.3 g. The mass of Na$_2$SO$_4$ is $1071.3 \times 0.08 = 85.7$ g. The mass of water is $1071.3 - 85.7 = 985.6$ g. The volume of 85.7 g of solid Na$_2$SO$_4$

is $85.7 \times 1000/2700 = 31.7$ cm^3. The volume of 985.6 g of water is $985.6 \times 1000/998.2 = 987.4$ cm^3. The total volume of the separate constituents is $31.7 + 987.4 = 1019.1$ cm^3. The contraction when the solution is made is $1019.1 - 1000 = 19.1$ cm^3. 1 dm^3 of the MgSO$_4$ solution has a mass of 1103.4 g. The mass of MgSO$_4$ is $1103.4 \times 0.1 = 110.3$ g. The mass of water is $1103.4 - 110.3 = 993.1$ g. The volume of 110.3 g of solid Na$_2$SO$_4$ is $110.3 \times 1000/2660 = 41.5$ cm^3. The volume of 993.1 g of water is $993.1 \times 1000/998.2 = 994.9$ cm^3. The total volume of the separate constituents is $41.5 + 994.9 = 1036.4$ cm^3. The contraction when the solution is made is $1036.4 - 1000 = 36.4$ cm^3. The two solutions have almost the same molar concentrations of ions: that of the Na$_2$SO$_4$ solution is $3 \times 0.602 = 1.806$ mol dm^{-3} and that of the MgSO$_4$ solution is $2 \times 0.917 = 1.834$ mol dm^{-3}. In terms of the molar concentration of ionic charge, the Na$_2$SO$_4$ solution has $2 \times 0.602 = 1.204$ mol dm^{-3} of positive charge and $2 \times 0.602 = 1.204$ mol dm^{-3} of negative charge. The MgSO$_4$ solution has $2 \times 0.917 = 1.834$ mol dm^{-3} of positive and negative charges. The greater molar charge concentration total is associated with the greater reduction in volume when the solutions are made. This is consistent with expectation that a greater charge would increase the attraction between the ions and water molecules. In this example it is clear that the effects of changing anions and cations and their charges cannot be separated.

1.3. The standard states of Na, water and dioxygen at 25 °C are solid, liquid and gas, respectively.

Chapter 2

2.1. The standard enthalpies of formation of the potassium ion and the cyanide ion are given in Tables 2.3 and 2.2, respectively. The enthalpy of solution of KCN is calculated from the enthalpies of formation of the products and reactants of the reaction:

$$KCN(s) \rightarrow K^+(aq) + CN^-(aq)$$
$$\Delta_{sol}H^\circ = \Delta_f H^\circ(CN^-, aq) + \Delta_f H^\circ(K^+, aq) - \Delta_f H^\circ(KCN, s)$$
$$= 150.6 - 252.4 + 113 = 11.2 \text{ kJmol}^{-1}$$

This is also given by the sum:

$$-\Delta_{latt}H^\circ(KCN, s) + \Delta_{hyd}H^\circ(K^+, g) + \Delta_{hyd}H^\circ(CN^-, g)$$
$$= 692 - 340 + \Delta_{hyd}H^\circ(CN^-, g)$$

$$\therefore 11.2 = 692 - 340 + \Delta_{\text{hyd}} H^\circ(\text{CN}^-, \text{g})$$
$$\therefore \Delta_{\text{hyd}} H^\circ(\text{CN}^-, \text{g}) = 11.2 - 692 + 340 = -341 \text{ kJ mol}^{-1}$$

2.2. The standard enthalpies of formation of the potassium ion and the hydroxide ion are given in Tables 2.3 and 2.2, respectively. The enthalpy of solution of KOH is calculated from the enthalpies of formation of the products and reactants of the reaction:

$$\text{KOH}(\text{s}) \rightarrow \text{K}^+(\text{aq}) + \text{OH}^-(\text{aq})$$
$$\Delta_{\text{sol}} H^\circ = \Delta_{\text{f}} H^\circ(\text{OH}^-, \text{aq}) + \Delta_{\text{f}} H^\circ(\text{K}^+, \text{aq}) - \Delta_{\text{f}} H^\circ(\text{KOH}, \text{s})$$
$$= -230 - 252.4 + 424.6 = -57.8 \text{ kJ mol}^{-1}.$$

This is also given by the sum:

$$-\Delta_{\text{latt}} H^\circ(\text{KOH}, \text{s}) + \Delta_{\text{hyd}} H^\circ(\text{K}^+, \text{g}) + \Delta_{\text{hyd}} H^\circ(\text{OH}^-, \text{g})$$
$$= 802 - 340 + \Delta_{\text{hyd}} H^\circ(\text{OH}^-, \text{g})$$
$$\therefore -57.8 = 802 - 340 + \Delta_{\text{hyd}} H^\circ(\text{OH}^-, \text{g})$$
$$\therefore \Delta_{\text{hyd}} H^\circ(\text{OH}, \text{g}) = -57.8 - 802 + 340 = -520 \text{ kJ mol}^{-1}$$

2.3. The standard enthalpies of formation of the sodium ion and the hydrogensulfate ion are given in Tables 2.3 and 2.2, respectively. The enthalpy of solution of NaHSO_4 is calculated from the enthalpies of formation of the products and reactants of the reaction:

$$\text{NaHSO}_4(\text{s}) \rightarrow \text{Na}^+(\text{aq}) + \text{HSO}_4^-(\text{aq})$$
$$\Delta_{\text{sol}} H^\circ = \Delta_{\text{f}} H^\circ(\text{HSO}_4^-, \text{aq}) + \Delta_{\text{f}} H^\circ(\text{Na}^+, \text{aq})$$
$$- \Delta_{\text{f}} H^\circ(\text{NaHSO}_4, \text{s})$$
$$= -887.3 - 240.1 + 1125.5 = -1.9 \text{ kJ mol}^{-1}$$

This is also given by the sum:

$$-\Delta_{\text{latt}} H^\circ(\text{NaHSO}_4, \text{s}) + \Delta_{\text{hyd}} H^\circ(\text{Na}^+, \text{g}) + \Delta_{\text{hyd}} H^\circ(\text{HSO}_4^-, \text{g})$$
$$= 784 - 424 + \Delta_{\text{hyd}} H^\circ(\text{HSO}_4^-, \text{g})$$
$$\therefore -1.9 = 784 - 424 + \Delta_{\text{hyd}} H^\circ(\text{HSO}_4^-, \text{g})$$
$$\therefore \Delta_{\text{hyd}} H^\circ(\text{HSO}_4^-, \text{g}) = -1.9 - 784 + 424 = -362 \text{ kJ mol}^{-1}$$

2.4. The standard enthalpies of formation of the sodium ion and the chlorate(VII) ion are given in Tables 2.3 and 2.2, respectively.

The enthalpy of solution of $NaClO_4$ is calculated from the enthalpies of formation of the products and reactants of the reaction:

$$NaClO_4(s) \rightarrow Na^+(aq) + ClO_4^-(aq)$$
$$\Delta_{sol}H^\circ = \Delta_fH^\circ(ClO_4^-,aq) + \Delta_fH^\circ(Na^+,aq) - \Delta_fH^\circ(NaClO_4,s)$$
$$= -129.3 - 240.1 + 383.3 = 13.9\,kJ\,mol^{-1}$$

This is also given by the sum:

$$-\Delta_{latt}H^\circ(NaClO_4,s) + \Delta_{hyd}H^\circ(Na^+,g) + \Delta_{hyd}H^\circ(ClO_4^-,g)$$
$$= 643 - 424 + \Delta_{hyd}H^\circ(ClO_4^-,g)$$
$$\therefore 13.9 = 643 - 424 + \Delta_{hyd}H^\circ(ClO_4^-,g)$$
$$\therefore \Delta_{hyd}H^\circ(ClO_4^-,g) = 13.9 - 643 + 424 = -205\,kJ\,mol^{-1}$$

2.5. The standard entropy of the Be^{2+} ion in the gas phase is given by Sackur–Tetrode as:

$$S^\circ(Be^{2+},g) = {}^3\!/_2\,R\ln 9 + 108.9 = 136.3\,J\,K^{-1}\,mol^{-1}$$

The standard entropy of the hydrated ion is $-129.8\,J\,K^{-1}\,mol^{-1}$ and the absolute entropy of hydration is given as:

$$\Delta_{hyd}S^\circ(Be^{2+},g) = -129.8 - 136.3 - 2\times 20.9 = -307.9\,J\,K^{-1}mol^{-1}$$

2.6. The standard entropy of the Sr^{2+} ion in the gas phase is given by Sackur–Tetrode as:

$$S^\circ(Sr^{2+},g) = {}^3\!/_2\,R\ln 87.6 + 108.9 = 164.7\,J\,K^{-1}mol^{-1}$$

The standard entropy of the hydrated ion is $-29.8\,J\,K^{-1}\,mol^{-1}$ and the absolute entropy of hydration is given as:

$$\Delta_{hyd}S^\circ(Sr^{2+},g) = -29.8 - 164.7 - 2\times 20.9 = -236.3\,J\,K^{-1}mol^{-1}$$

Chapter 3

3.1. CsCl: mass per $dm^3 = 1925 \times 0.6564 = 1260$ g; molar concentration $= 1260/168.4 = 7.48\,mol\,dm^{-3}$. $BaCl_2$: mass per $dm^3 = 1292 \times 0.2698 = 348$ g; molar concentration $= 348/208.3 = 1.67\,mol\,dm^{-3}$. $BaSO_4$: mass per $dm^3 = 1000 \times 0.0000024 = 0.0024$ g;

molar concentration $= 0.0024/233.3 = 1.03 \times 10^{-5}$ mol dm^{-3}. Increasing the cation charge from $+1$ to $+2$ decreases the solubility, and if both cation and anion have double charges the solubility is very slight.

3.2. (a) Nothing happens; the least soluble combination is K_2SO_4 but it is soluble to the extent of 0.69 molal $[2 \times 0.69 = 1.38$ molal in K^+(aq)] and the molality of K^+(aq) in the mixed solution is only 1.0 molal. (b) $MgSO_4$ is precipitated; it is the least soluble ion combination. For the same reason the compounds precipitated in (c), (d), (e) and (f) are $CaCO_3$, $MgCO_3$, $CaSO_4$ and $MgCO_3$, respectively.

3.3. $KClO_4$ is considerably less soluble than $NaClO_4$ because of the larger K^+ ion; $KClO_4$ has a more negative lattice enthalpy and the smaller Na^+ has a more negative enthalpy of hydration.

3.4. In solution the Cs^+(aq) and F^-(aq) ions are both 37.4 molal. The molality of water is $1000/18.015 = 55.5$ molal. Each ion would have a share of water molecules of $55.5/(2 \times 37.4) = 0.74$. This remarkable result emphasizes the difficulties of applying formal thermodynamics to such concentrated solutions. There is insufficient water present to hydrate the ions fully as they would be in a dilute solution.

3.5. For hydroxide formation:

$$M^{4+}(aq) + H_2O(l) \rightarrow MOH^{3+}(aq) + H^+(aq)$$
$$MOH^{3+}(aq) + H_2O(l) \rightarrow M(OH)_2^{2+}(aq) + H^+(aq)$$
$$M(OH)_2^{2+}(aq) + H_2O(l) \rightarrow M(OH)_3^+(aq) + H^+(aq)$$
$$M(OH)_3^+(aq) + H_2O(l) \rightarrow M(OH)_4(s) + H^+(aq)$$

For oxide formation:

$$M(OH)_2^{2+}(aq) \rightarrow MO_2(s) + 2H^+(aq)$$

Chapter 4

4.1. (a) $Pt^{2+}(aq) + Sn(s) \rightarrow Sn^{2+}(aq) + Pt(s)$; $E^\circ = +1.18 - (-0.14) = +1.32$ V, the reaction is feasible. (b) $Co^{2+}(aq) + V(s) \rightarrow V^{2+}(aq) + Co(s)$; $E^\circ = -0.28 - (-1.18) = +0.9$, the reaction is feasible. Or $3Co^{2+}(aq) + 2V(s) \rightarrow 2V^{3+}(aq) + 3Co(s)$; $E^\circ = -0.28 - (-0.87) = +0.59$, the reaction is also feasible. (c) $Mg^{2+}(aq) + Pb \rightarrow Pb^{2+}(aq) + Mg(s)$; $E^\circ = -2.37 - (-0.13) = -2.24$ V, the reaction is not feasible. (d) $Ba^{2+}(aq) + Cd(s) \rightarrow Cd^{2+}(aq) + Ba(s)$; $E^\circ = -2.91 - (-0.4) = -2.51$ V, the reaction is not feasible.

4.2. (a) $E^\circ(Mn^{3+}/Mn^{2+}) = 3E^\circ(Mn^{3+}/Mn) - 2E^\circ(Mn^{2+}/Mn) = (-0.28 \times 3) - (2 \times -1.19) = +1.54$ V. (b) $E^\circ(V^{3+}/V^{2+}) = 3E^\circ(V^{3+}/V) - 2E^\circ(V^{2+}/V) = (-0.87 \times 3) - (2 \times -1.18) = -0.25$ V. (c) $E^\circ(Fe^{3+}/Fe^{2+}) = 3E^\circ(Fe^{3+}/Fe) - 2E^\circ(Fe^{2+}/Fe) = (-0.04 \times 3) - (2 \times -0.45) = +0.78$ V.

4.3. $HOCl(aq) + H^+(aq) + e^- \rightleftharpoons \frac{1}{2}Cl_2(g) + H_2O(l)$
$$E^\circ = +1.61 \text{ V}$$

$\frac{1}{2}Cl_2(g) + e^- \rightleftharpoons Cl^-(g)$
$$E^\circ = +1.36 \text{ V}$$

adding: $HOCl(aq) + H^+(aq) + 2e^- \rightleftharpoons Cl^-(g) + H_2O(l)$
$$E^\circ = (1.61 + 1.36)/2 = +1.49 \text{ V}$$

reversing: $Cl^-(g) + H_2O(l) \rightleftharpoons HOCl(aq) + H^+(aq) + 2e^-$
$$E^\circ = -1.49 \text{ V}$$

$ClO_4^-(aq) + 8H^+(aq) + 8e^- \rightleftharpoons Cl^-(g) + 4H_2O(l)$
$$E^\circ = 1.39 \text{ V}$$

adding: $ClO_4^-(aq) + 7H^+(aq) + 6e^- \rightleftharpoons HOCl(aq) + 3H_2O(l)$
$$E^\circ = [(2 \times -1.49) + (8 \times 1.39)]/6 = 1.36 \text{ V}$$

4.4. $H_2O_2(aq) + 2H^+(aq) + 2e^-(g) \rightleftharpoons 2H_2O(l)$ $\qquad E^\circ = 1.78$ V
$2e^-(aq) \rightleftharpoons 2e^-(g)$ $\qquad E^\circ = -1.85$ V
$H_2O_2(aq) + 2H^+(aq) + 2e^-(aq) \rightleftharpoons 2H_2O(l)$
$$E^\circ = [(2 \times 1.78) + (2 \times -1.85)]/2 = -0.07 \text{ V}$$

On the e^- (aq) scale the values of standard reduction potentials are more negative than those on the conventional scale.

Chapter 5

5.1. (i) $H_3AsO_4(aq) + 2H^+(aq) + 2e^- \rightleftharpoons HAsO_2(aq) + 2H_2O(l)$
(ii) $HBrO(aq) + H^+(aq) + 2e^- \rightleftharpoons Br^-(aq) + H_2O(l)$
(iii) $HOClO(aq) + 3H^+(aq) + 4e^- \rightleftharpoons Cl^-(aq) + 2H_2O(l)$
(iv) $MnO_4^-(aq) + 4H^+(aq) + 3e^- \rightleftharpoons MnO_2 + 2H_2O(l)$

5.2. (i) $E = +0.56 - 0.0592$ pH
(ii) $E = +1.33 - 0.0296$ pH
(iii) $E = +1.57 - 0.0444$ pH
(iv) $E = +1.68 - 0.0789$ pH

5.3. $E^\circ(M^{2+}/M) = [E^\circ(M^{2+}/M^+) + E^\circ(M^+/M)]/2 = (-0.2 + 0.6)/2 = 0.4/2 = 0.2$ V. M^+ disproportionates.

5.4. Trioxygen has the potential to oxidize water to dioxygen since the reduction potential has a value greater than $1.23 + 0.41 = 1.64$ V, the practical limit for such a process. In solution at pH $= 0$, O_3 is unstable. Dioxygen, itself a powerful oxidizing agent, is stable in solution at pH $= 0$ only if there are no reducing agents present. At pH $= 14$, the sulfate(VI) is very stable. Sulfate(IV) ion is a good reducing agent, but will not reduce water; it is within the practical limit.

5.5. The two half-reactions are:

$$Cr^{3+} + e^- \rightleftharpoons Cr^{2+} (E^\circ = -0.41 \text{ V})$$
$$Zn \rightleftharpoons Zn^{2+} + 2e^- (E^\circ = +0.76 \text{ V})$$

The overall reaction is:

$$Zn + 2Cr^{3+} \rightleftharpoons Zn^{2+} + 2Cr^{2+} \quad [E^\circ = 0.76 - 0.41 = +0.35 \text{ V}]$$

which is feasible (no multiplying of E° values allowed!). The reduction potential for the Cr^{3+}/Cr^{2+} is on the borderline for stability of the lower oxidation state reacting with water to give dihydrogen, but a more serious threat to the stability of Cr^{2+} is the presence of dissolved dioxygen, which has the potential to re-oxidize the ion to the $+3$ state.

5.6. The half-reaction is:

$$Cr_2O_7^{2-} + 14H^+ + 6e^- \rightleftharpoons 2Cr^{3+} + 7H_2O \quad (E^\circ = +1.33 \text{ V})$$

The appropriate Nernst equation is:

$$E = 1.33 - (2.303 \times 14 \, RT/6F) \text{ pH}$$
$$= 1.33 - (2.303 \times 14 \, RT/6F)[\text{pH} = 1] = +1.19 \text{ V}$$

Chapter 6

6.1. The apparent oxidation state of Tl in TlI_3 is $+3$, based on the idea that the oxidation state of the more electronegative iodine atom is -1 as expected for an octet configuration. However, the mixture of $Tl^{3+}(aq)$ and $I^-(aq)$ is likely to cause the oxidation of iodide ion to iodine and the resulting compound is one of Tl^I, *i.e.* $Tl^+(I_3)^-$.

6.2. The equation for the calculation of the enthalpy change for the reaction:

$$\tfrac{1}{2}X_2(g,\ l\ or\ s) + \tfrac{1}{2}H_2(g) \rightleftharpoons H^+(aq) + X^-(aq)$$

is:

$$\text{Reduction enthalpy} = \Delta_a H^\circ(X, g) + E(X) + \Delta H^\circ(X^-) + 420$$

where $E(X)$ is the electron attachment enthalpy of the gaseous X atom, the 420 amount being the "oxidation enthalpy" for the standard reference reaction (the reverse of reaction 6.1). The E° values are obtained by dividing the reduction enthalpies by $-F$. The results are:

	F	Cl	Br	I
E° (calc)/V	+3.45	+1.73	+1.25	+0.57

The discrepancies between the observed and calculated values for E° are attributable to the missing entropy terms, but are not large.

6.3. Br^V does not have the power to oxidize bromide ion beyond the elementary stage. The reaction with bromide ion is:

$$BrO_3^-(aq) + 5Br^-(aq) + 6H^+(aq) \rightarrow 3Br_2(aq) + 3H_2O(l)$$

The potential for the reaction is $[(+1.45 \times 4) + 1.6]/5 - 1.09 = +0.39$ V.

6.4. The standard reduction potentials for the $+7/-1$ reductions for ClO_4^-, BrO_4^- and H_5IO_6 to their -1 ions are given by:

$$ClO_4^-/Cl^- : [(1.19\times2) + (1.21\times2) + (1.65\times2) + 1.61 + 1.36]/8 = 1.38 \text{ V}$$
$$BrO_4^-/Br^- : [(1.85\times2) + (1.45\times4) + 1.6 + 1.09]/8 = 1.52 \text{ V}$$
$$H_5IO_6/I^- : [(1.6\times2) + (1.13\times4) + 1.44 + 0.54]/8 = 1.21 \text{ V}$$

The BrO_4^- ion is the most powerful oxidizing agent.

6.5. Three thermochemical cycles are required for the solution of the problem. (i) One (not shown, but similar to that in Figure 2.11) is for the formation of $NH_4^+(aq)$ and $ClO_4^-(aq)$ from the solid compound, and the splitting of the lattice into the gaseous ions followed by their hydration. This leads to the equations:

$$\Delta_{sol}H^\circ(NH_4ClO_4) = -132.5 - 129.3 - (-295.3) = +33.5\,kJ\,mol^{-1}$$

$$\therefore 33.5 = 586 + \Delta_{hyd}H^\circ(NH_4^+, g) - 205$$

$$\therefore \Delta_{hyd}H^\circ(NH_4^+, g) = 33.5 - 586 + 205 = -347.5\,kJ\,mol^{-1}$$

The second cycle (again similar to that of Figure 2.11) is for the formation of $2NH_4^+(aq)$ and $SO_4^{2-}(aq)$ from the solid compound, and the splitting of the lattice into the gaseous ions followed by their hydration. This leads to the equations:

$$\Delta_{sol}H^\circ[(NH_4)_2SO_4] = -(2 \times 132.5) - 909.3 - (-1180.9)$$
$$= +6.6\,kJ\,mol^{-1}$$

$$\therefore 6.6 = 1789 + 2\Delta_{hyd}H^\circ(NH_4^+, g) - 1099$$

$$\therefore \Delta_{hyd}H^\circ(NH_4^+, g) = (6.6 - 1789 + 1099)/2 = -342.7\,kJ\,mol^{-1}.$$

The mean of the two values of $\Delta_{hyd}H^\circ(NH_4^+, g) = -345\,kJ\,mol^{-1}$.
The third cycle is:

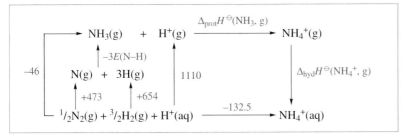

(ii) From the left side of the cycle:
$-46 = 473 + 654 - 3E(N-H)$; $\therefore E(N-H) = (473 + 654 + 46)/3 = 391\,kJ\,mol^{-1}$, an estimate of the bond energy term for the N–H bond in ammonia.

(iii) From the remainder of the cycle:

$$-132.5 = -46 + \Delta_{prot}H^\circ(NH_3, g) - 345 + 1110$$

$$\therefore \Delta_{prot}H^\circ(NH_3, g) = -132.5 + 46 + 345 - 1110 = -852\,kJ\,mol^{-1}.$$

(iv) For the ammonium ion: $E(N-H) = [(3 \times 391) + 852]/4 = 506\,kJ\,mol^{-1}$

Chapter 7

7.1. The estimate of $\Delta_{hyd}H^\circ(Cr^{2+}, g)$ is given by the equation:

$$\Delta_{hyd}H^\circ(Cr^{2+}, g) = \Delta_f H^\circ(Cr^{2+}, aq) - \Delta_a H^\circ(Cr, g)$$
$$+ 2\Delta_{hyd}H^\circ(H^+, g) + 2I_H - I_1(Cr) - I_2(Cr)$$
$$+ 2\Delta_a H^\circ(H, g)$$
$$= -143.5 - 397 - (2 \times 1110) + 2 \times 1312 - 653$$
$$- 1591 + (2 \times 218)$$
$$= -1945 \, kJ \, mol^{-1}$$

7.2. The estimate of $\Delta_{hyd}H^\circ(Cu^+, g)$ is given by the equation:

$$\Delta_{hyd}H^\circ(Cu^+, g) = \Delta_f H^\circ(Cu^+, aq) - \Delta_a H^\circ(Cu, g) + \Delta_{hyd}H^\circ(H^+, g)$$
$$+ I_H - I_1(Cu) + \Delta_a H^\circ(H, g)$$
$$= 71.7 - 337 - 1110 + 1312 - 745 + 218$$
$$= -590 \, kJ \, mol^{-1}$$

7.3. The estimate of $\Delta_{hyd}H^\circ(Sc^{3+}, g)$ is given by the equation:

$$\Delta_{hyd}H^\circ(Sc^{3+}, g) = \Delta_f H^\circ(Sc^{3+}, aq) - \Delta_a H^\circ(Sc, g)$$
$$+ 3\Delta_{hyd}H^\circ(H^+, g) + 3I_H - I_1(Sc) - I_2(Sc)$$
$$- I_3(Sc) + 3\Delta_a H^\circ(H, g)$$
$$= -614.2 - 378 - (3 \times 1110) + (3 \times 1312) - 633$$
$$- 1235 - 2389 + (3 \times 218)$$
$$= -3989 \, kJ \, mol^{-1}$$

7.4. (i) Co^{3+}; (ii) Zn^{2+}; (iii) Co^{2+}; (iv) Zn; (v) Co^{3+} and Ag^+.

7.5. The overall reaction potential is $+0.44 - 0.257 = +0.183$ V. The reaction is feasible in the direction written in the equation. The calculated value for E° is $+34/2F = +0.176$ V.

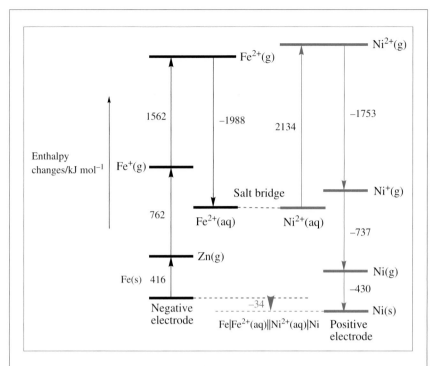

7.6. TiO^{2+}, VO_2^+, Cr^{3+}, Mn^{2+}, Fe^{3+} and Co^{2+} are most stable in the presence of dioxygen at pH $= 0$.

Chapter 8

8.1. The estimate of $\Delta_{hyd}H^{\circ}(Ho^{3+}, g)$ is given by the equation:

$$\begin{aligned}
\Delta_{hyd}H^{\circ}(Ho^{3+}, g) &= \Delta_f H^{\circ}(Ho^{3+}, aq) - \Delta_a H^{\circ}(Ho, g) \\
&\quad + 3\Delta_{hyd}H^{\circ}(H^+, g) + 3I_H - I_1(Ho) - I_2(Ho) \\
&\quad - I_3(Ho) + 3\Delta_a H^{\circ}(H, g) \\
&= -707 - 300.8 - (3 \times 1110) + (3 \times 1312) - 581 \\
&\quad - 1139 - 2204 + (3 \times 218) \\
&= -3672 \text{ kJ mol}^{-1}
\end{aligned}$$

8.2. Using equation (8.2):

$$\begin{aligned}
E^{\circ}(Lu^{3+}/Lu) &= -[-\Delta_{hyd}H^{\circ}(Lu^{3+}) - I_3(Lu) - I_2(Lu) - I_1(Lu) \\
&\quad - \Delta_a H^{\circ}(Lu) + (3 \times 420)] \div 3F = -[3758 - 2022 - 1341 - 524 - 428 \\
&\quad + 1260] \div 3F = -2.43 \text{ V}
\end{aligned}$$

8.3. The table gives the required data (kJ mol^{-1} for enthalpies and ionization energies) and the calculated values of the standard reduction potentials for the three elements.

	$\Delta_aH°$	I_1	I_2	I_3	$\Delta_{hyd}H°$	$E°$ (calc.)/V	$E°$ (obs.)/V
Cs	+77	376			−291	−2.67	−3.03
Ba	+180	503	965		−1346	−2.79	−2.91
La	+431	538	1067	1850	−3335	−2.45	−2.38

Barium is almost as powerful a reducing agent as is Cs, but La is somewhat less so. The appropriate sums of the ionization energies and the enthalpies of atomization compete well with the enthalpies of hydration for Cs and Ba, but the enthalpy of hydration of the La^{3+} ion is not sufficiently negative to outweigh the large third ionization energy.

Subject Index